School Choice and School Governance

A Historical Study of the United States and Germany

Jurgen Herbst

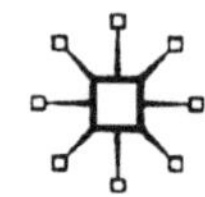

SCHOOL CHOICE AND SCHOOL GOVERNANCE

First published in 2006 by
PALGRAVE MACMILLAN™
175 Fifth Avenue, New York, N.Y. 10010 and
Houndmills, Basingstoke, Hampshire, England RG21 6XS
Companies and representatives throughout the world.

PALGRAVE MACMILLAN is the global academic imprint of the Palgrave Macmillan division of St. Martin's Press, LLC and of Palgrave Macmillan Ltd. Macmillan® is a registered trademark in the United States, United Kingdom and other countries. Palgrave is a registered trademark in the European Union and other countries.

ISBN 1–4039–7302–4

Library of Congress Cataloging-in-Publication Data

Herbst, Jurgen.
School choice and school governance : a historical study of the United States and Germany / Jurgen Herbst.
p. cm.
Includes bibliographical references and index.
ISBN 1–4039–7302–4
1. School choice—United States—History. 2. School choice—Germany—History. 3. School management and organization—United States—History. 4. School management and organization—Germany—History. I. Title.

LB1027.9.H485 2006
379.1′110973—dc22 2005056614

A catalogue record for this book is available from the British Library.

Design by Newgen Imaging Systems (P) Ltd., Chennai, India.

First edition: April 2006

10 9 8 7 6 5 4 3 2 1

Printed in the United States of America.

For Sheyanne

Contents

List of Tables ix

Preface and Acknowledgments xi

Introduction School Choice: A Brief Overview 1

1 Beginnings 7

2 The Systematization of Public Education 31

3 School Governance and School Choice 1900–1950 63

4 School Choice in the United States after World War II 95

5 *Schulwahl* in the Post–World War II Period 121

6 Retrospect and Outlook 143

Notes 163

Name Index 189

Subject Index 193

List of Tables

2.1 Number of Public Elementary Schools in Prussia, 1861–1901 55

2.2 Number and Percentage of Jewish Students in Prussia's Christian and Jewish Public Elementary Schools, 1886–1901 55

3.1 Enrollment Totals and Percentages of U.S. Public and Private High and Elementary Schools, 1900–1950 73

3.2 Number and Percentage of U.S. Private Secular and Denominational High Schools and Their Students,1900–1915 74

3.3 Comparative Data Concerning U.S. Public and Private High Schools 1900–1914 74

3.4 Number of U.S. Church-Sponsored High Schools and Their Students, 1900–1917. Major Denominations only 75

3.5 Establishment of U.S. Private Schools by Time Period. Before 1904, 1904–1953, 1954–1993 in percent of those in existence 1993/1994 76

3.6 Public Elementary Schools in Prussia, 1921–1932 84

4.1 Enrollments in Thousands in U.S. Public and Private Schools from Kindergarten to Grade Twelve with Percentages of Enrollments in Private Institutions 1950–2000 111

4.2 Number and Percentage Distribution of U.S. Public and Private School Students by Grade Level, 1999–2000 112

4.3 Number of Private Schools and Students by Category, Fall 1999 113
4.4 Number and Percentage of Students in U.S. Private Schools by Category and Grade Level, Fall 1999 114
5.1 Socioeconomic Composition of Students in German Schools in 2000 139
5.2 Comparison of Immigrant Families, Germany and United States in 2000 141

Preface and Acknowledgments

In the United States school choice is today a subject of debate. It began to make headlines during the 1950s as it was viewed by many parents and legislators as one answer to what was seen as the crisis of public schooling: A deterioration of facilities and of student discipline in inner city schools, a perceived lessening of academic standards nationwide, and a simultaneous demand that public schools become the engines of national progress in all areas of economic, scientific, and cultural life. In several Southern states the 1954 desegregation decision of the U.S. Supreme Court in *Brown v. Board of Education* strengthened agitation for parental choice as a means of escaping integrated public schools. When the cold war with the Soviet Union heated up and Congress passed the National Defense Education Act in 1958 both academic proponents of strengthened science education in the schools and free-wheeling lay pedagogical reformers endorsed school choice proposals of various kinds. By the 1960s free market advocates as well as liberal scholars introduced the idea of vouchers as favored means for financing school choice programs. School choice had become a front-page subject.

In the reunified Germany of the 1990s and the early years of the twenty-first century, school choice had assumed two different meanings. As the country's educational system, administered in the *Länder* of the Federal Republic, had returned to its traditional dual system of academic and vocational education and its three-pronged arrangement of elementary, middle, and higher schools, school choice known as *Schulwahl* referred to the selection of an advanced school for those children who, at age ten, were thought capable of a secondary education. This was a practice that had come into use during the second half of the nineteenth century. Today, however, the meaning of the term *Schulwahl* also includes parental participation in a great variety of choices made concerning the education of their children. These choices may pertain to the level of schooling but can also affect any other educational issue that in the past had been customarily left to the authorities of Germany's state-directed and controlled public school system.

In this book I investigate the antecedents of today's school choice policies in the United States and Germany. I do not attempt here to offer a full-fledged comparative history, but think of my presentation as an exercise in contrasting history—a history that gives background and contrast to its main subject (American school choice) by setting it before or against a related, though not identical, subject (German *Schulwahl*). In our age of worldwide interrelatedness one cannot present adequately the history of one country's school choice if one makes that subject the single item of observation. When one does that, one can view and understand it only through its own genesis and development in the context of its native environment. When, however, one sees it against the background of a similar item in a different setting, a fuller, more encompassing perspective becomes possible. Thus my focus in this book has remained throughout on the United States with Prussia and Germany serving as background for the American story.

I should also alert my readers to my realization that the subject of school choice is inextricably intertwined with the subject of school governance. The one cannot be discussed without the other, and both together present a general history of public education policy. As in this book I look at the interactions of parents with school authorities and classroom teachers, I also review the public debates over legislative and judicial issues concerning the public schools.

I address this book primarily to a readership of citizens who, like I myself, seek to gain a better understanding of public events. My participation as a lecturer in the Lifelong Learning Program of Fort Lewis College in Durango, Colorado, has immensely increased my appreciation of such general town-gown audiences of interested lay persons, teachers, school administrators, state legislators, businessmen, and businesswomen, professionals of many kind, college faculty members and college students. Their reactions and questions have been an inspiration. Writing this book, I have had them in mind. But college instructors, too, will find the book a helpful text for both graduate and undergraduate classes in educational history and policy studies.

Earlier publications drawing on some of the material used in the preparation of this book include my "Schools Between State and Civil Society in Germany and the United States—An Historical Perspective," which appeared in Heinz-Dieter Meyer and William Lowe Boyd, eds., *Education between State, Markets, and Civil Society: Comparative Perspectives* (Mahwah, NJ: Lawrence Erlbaum Associates, 2000), and my "Nineteenth Century Schools Between Community and State: The Cases of Prussia and the United States," in the *History of Education Quarterly*, 42 (Fall 2002): 317–341.

It is now my pleasant duty to acknowledge the support I received during the many months I spent gathering material for this study and putting my analyses and conclusions into final form. My thanks go to the Spencer Foundation whose generous financial support enabled me to carry out research in German archives. I am especially indebted to Professor Manfred Heinemann of the University of Hannover whose hospitality at his Zentrum für Zeitgeschichte von Bildung und Wissenschaft and whose sage advice proved to be of inestimable value. I also gladly acknowledge the assistance I received at the Wissenschaftszentrum in Bonn, from Dr. Renate Martini at the Deutsches Institut für Internationale Pädagogische Forschung in Frankfurt/Main, and from Professor Frank-Rüdiger Jach in Hannover. Advice, encouragement, and support also have come from Professor William J. Reese of the University of Wisconsin, Madison, and Professor Gary McCulloch of the University of London's Institute of Education. As much as I have relied on their advice and counsel, I cannot claim that they necessarily agree with the views and opinions I express in these pages. I alone am responsible for them and for any errors of fact and interpretation found in this book. My thanks also go to Phyllis Kroupa and her staff at the Interlibrary Loan Department of the Reed Library at Fort Lewis College. Without their help I would never have been able to carry this study to a conclusion. And as always there has been Sue, my life's love, who listened and counseled as only she can do.

Jurgen Herbst
Durango, Colorado

Introduction

School Choice: A Brief Overview

School choice has become an often heard phrase in public debates and private conversations. As a concept that refers to the desire of parents to send their children to schools they and not outside educational or governmental authorities have selected, it is easily understood. In European countries and the United States as well as in Australia and New Zealand, citizens dissatisfied for various reasons with the performance of their public schools, have begun to ask for the right and the opportunity to send their children to public or private schools of their choice.[1]

If we ask today what has spurred on these parental demands, the answers we receive are varied. They range from dissatisfaction with the academic standards of the school their children attend to objections to the skin color, social class, language, or gender of their children's schoolmates. They express a desire to have their children exposed to a different teaching style or complain of the administrative or bureaucratic atmosphere that characterizes their children's school. They point to specific subjects taught in school—usually subjects that carry moral overtones—or to the absence or specific kind of religious instruction offered.

Of these answers the one that points to classroom religious instruction or its absence has fueled debates and struggles over schooling ever since the sixteenth century. In Europe, schools offered religious instruction according to the creed and practices of the state's established church. While in today's Germany both secular and religious schools exist, in public schools associated with one of the major denominations—Protestant-Lutheran, Protestant-Reformed, or Roman Catholic—denominational religious instruction is a recognized and, with few exceptions for Jewish, Muslim, or agnostic children, a required subject. In communities in which secular and confessional public schools exist, parents are free to choose the type of school their children will attend.

In the United States the First Amendment to the U.S. Constitution has led to a quite different situation. In its establishment clause it declares that "Congress shall make no law respecting an establishment of religion," and in its free exercise clause it adds that it also may not prohibit the free exercise of religion. The establishment clause has often been held to require the separation of state and church and to forbid the teaching and funding of religion in public schools. Religion, therefore, is to be taught only in churches and private schools. Some school choice advocates, however, argue that the free exercise clause of the amendment requires an opposed interpretation. They ask that students or their parents be publicly funded through vouchers to attend religious schools. Other school choice proponents seek to avoid this conundrum by suggesting that religious schools participating in a publicly funded choice plan keep sectarian functions separate from educational ones and be chartered as public schools. They could then receive public funding directly from federal, state, or district sources rather than indirectly through parents or students.[2] Whether religious schools would comply with the demand for keeping sectarian religion out of the classroom remains a question. The Supreme Court decision in *Zelman v. Simmons-Harris*, 536 U.S. 639 (2002), ruling voucher programs constitutional, may have made that question irrelevant.

Though it is true that in Germany as well as in the United States the existence of legally guaranteed religious and secular private schools has provided a measure of school choice to parents who want to avail themselves of that opportunity, in many instances this freedom has been circumscribed or is nonexistent. The requirement of paying tuition and other fees or the pressures often exerted by religious authorities on parents to send their children to the school of their denomination or church have placed limitations on the parental freedom of choice.

Parents have met resistance to their desires for school choice even in the public schools of both countries. Unless there were or are specific reasons concerning a child's health or mental ability that in the United States bring into play the provisions of the 1975 Public Law 94-142, the "Education for all Handicapped Children Act,"[3] state-administered public school systems in both countries have been regulating attendance in elementary schools by geographic location.[4] In German public secondary education the results of academic ability and manual skill testing nearly always determine the high school or vocational school a child will attend. In American secondary education the consolidation of rural high schools in the late nineteenth and early twentieth centuries rarely considered parental preferences. It left parents with no choice but to send their children to the consolidated school or to deny them a high school education altogether. For African-American parents the history of legally enforced racial segregation is a potent reminder of the denial of choice. This practice came to an end only with the 1954 Supreme Court decision in *Brown v. Board of Education*. In the following decades court-ordered busing, seeking to overcome the effects of racial separation through integration, also largely ignored or overrode parental preferences. For advocates of school choice its past has not been encouraging and, at the same time, largely unexplored. Its current implications for the future of public and private education are only dimly perceived and shrouded in controversy. A better understanding of its history should help us to evaluate its present and future prospects.

American historians of education also have done little to cast light on the history of school choice and private schools. Employed as most of them are in teacher training programs in state universities and colleges, it seems natural and perhaps unavoidable that they have traditionally been committed to public schools and their teachers, and have found few occasions to consider the history of private schools or to look upon the subject of schooling from the point of view of parents.[5] Though Bernard Bailyn reminded them in the 1960s that education was not limited to public schooling and Lawrence Cremin challenged them to investigate other areas, such as the history of the family, of women, of libraries, of news media, summer camps, adult education, and more, the history of private education and school choice remains a virtual terra incognita.[6]

As a result there exists hardly a textbook in the United States that devotes as much space to the history of private and parochial schools and their problems as it does to the history of the public school. With the notable exception of Merle Curti's *The Social Ideas of American Educators*, hardly any of the most commonly used texts devotes more than a few perfunctory lines to the place and significance of Bishop Spalding and the Catholic school system in American education. I know of only few investigations that deal with the historical antecedents of such recent innovations as vouchers, charter and magnet schools, or with the development of such a longstanding tradition as local or community control.[7] It seems to be in bad taste to call for textbook treatment of such issues as privatization, market forces in education, and home schooling, unless they are quickly disposed of as being beyond the pale of legitimate educational topics. The most one can find are references to the past and present existence of parochial schools and denominational colleges and universities.

It is also puzzling that the American literature on progressive education has paid little attention to the role of parents and of school choice. Partly responsible for this is Lawrence Cremin's 1961 *The Transformation of the School: Progressivism in American Education, 1876–1957*. This book exerted a dominant influence over the historiography of American education and, far from recognizing the place of educational progressivism in the history of pedagogy, defined it as the educational side of an American political reform movement called Progressivism. But progressive education, known as *Reformpädagogik* in German and *éducation nouvelle* in French, carries a long tradition of attempting to make room for parent, teacher, and student initiatives in public schools and to establish nonpublic schools on such a basis.[8] Such efforts, diverse as they have been in their details, have one thing in common. They protest the uniformity and inflexibility of officially established school systems that, in the reformers' view, discipline and infantilize children in order to prepare them for adulthood. They reject the "snare of preparation" and instead seek to affirm childhood as a stage of life in its own right.

From the beginning of our established public school systems such reform efforts have been wedded to public schooling like the "yin" to the "yang." The one could not exist without the other.[9] School choice is merely the latest of such attempts at school reform.

In Germany the dominance of public schools dates back to the beginning of the eighteenth century when they were considered in Prussian law to be "institutions of the state." This is true today as much as it was then. The Max Planck Institute authors of the 1990 study of the German educational system stated flatly: "There can be no question, *the* school of the Federal Republic is a public school [*Staatsschule*]."[10] Frank-Rüdiger Jach wrote the following nine years later: In the Federal Republic "state supervision of schooling is widely understood as the essence of the state's governing rights in matters of organization, planning, direction and supervision of schools."[11] Even private schools are considered to belong in the public sphere and are placed under the joint administrative and regulative authority of state and federal governments. They stand outside of or in opposition to the mainstream of German educational history.

One cannot help but also note that in Germany the long tradition of state supervision of schools has only recently prompted some historians of education to question the compatibility of an education delivered and directed from above with the notion of an education for democracy and freedom. As Hellmut Becker, one of Germany's outstanding educators, wrote in 1954: The German school has become an

> administrated school. . . . It stands on the lowest level of the state's administrative hierarchy, . . . comparable to a local internal revenue or employment office or the local police precinct, in clear contrast to the self-governing authority of city hall. Its teachers have become functionaries; and the school is in danger of training functionaries only.[12]

While Becker's words may overstate the issue, they nonetheless help us to understand why parents and advocates of school choice have not always found open ears in Germany's school bureaucracy.

The following chapters seek to outline the development of American and German theories, ideas, proposals, and practices, which may be considered antecedents of current attempts to find a greater and more effective role for parents in their children's schools. They will include proposals and discussions that concern the parental desire to have a say in the formal deliberations on school choice as well as parental demands to gain a hearing on matters of their children's schools' curricula, discipline, and daily procedures. It is my hope that these pages may permit the reader to gain a deeper and historically grounded understanding of today's school choice debates.

Readers may also ask why I selected Prussian and German developments of school choice as the background for my discussion of the American story. The reason is primarily historical and focuses on the many instances of mutual influences. In the nineteenth century the most pronounced foreign influence on modern American public education was that of Prussia. During the 1820s and 1830s many reports on Prussian schools and their practices written by American travelers reached the United States and spurred on what came to be known as the Public School Revival. Widely discussed by American educators was the Frenchman Victor Cousin's *Report on the State of Public Instruction in Prussia*, published in New York in 1835.[13] Eighty years later it was the Munich, Germany, school superintendent Georg Kerschensteiner whose practices and lectures heavily influenced the American debate over the introduction and control of vocational education. The years after 1945 then witnessed a reverse movement when under the banner of reeducation many American educators contributed significantly to developments in public education in Germany. The educational discourse between German and American educators has been frequent and influential enough, I believe, to justify my selection of the German experience of school choice as a background to the American story.[14]

Chapter 1

Beginnings

Fresh and contemporary as the term school choice appears to us today, it has a history that in the English-speaking world leads back to Adam Smith's 1776 *Wealth of Nations*, to Thomas Paine's 1792 *Rights of Man*, and in the nineteenth century to John Stuart Mill's *On Liberty*. The German language's most famous manifesto was Wilhelm von Humboldt's "Ideas for an Attempt to Determine the Limits of the State's Effectiveness," first published in 1792. While not explicitly using the term school choice, the essay speaks of the need to protect a people's education from state control. In all of these instances the growing presence of modern state administrations had provoked a debate over the provision of schooling for a state's or country's children.

Before that happened, schooling of children was undertaken by many different agencies ranging from the family and churches, local community leaders and organizations to provincial or state governments. There was no uniform overall arrangement. Government, in its role as protector and supporter of the public welfare, looked upon its educational responsibilities mainly as a matter of economic requirements and social discipline. More often than not, it left schooling to be funded by townships and municipalities or expected churches, private schoolmasters, and employers of youth to provide it. Parents were relatively free to avail themselves of or to ignore the educational opportunities offered to their children. Parental school choice, while sometimes possible depending on local opportunities, was not a relevant issue.

In the United States, school choice as an issue entered the public debate when during and after the Revolution voices were raised to demand statewide or national schooling. Until then the ethnic and religious diversity of colonial America had made schooling primarily a matter of local endeavor, though Massachusetts had experienced an earlier conflict during the middle decades of the eighteenth century. Merchants and other businessmen, favoring a practical English education for their town's children, had protested the General Court's attempts to enforce the establishment of municipal Latin grammar schools.[1] Schooling, they had held, was a matter best left to the discretion of the localities that supported it and were better informed than a central government about their communities' economic and educational needs. But during the Revolution's upheavals more general concerns came to the fore, and several influential Revolutionary leaders argued that neighborhood dame schools, private schoolmasters, apprenticeship arrangements, and church and community efforts for various types of academies and Latin schools were no longer sufficient. Educators, they held, had to embrace state- and nationwide perspectives. With that the debate began over the place of education in the nation's affairs, a debate that would eventually feed into the agitation over parental school choice.

In Virginia, where the members of the widely dispersed white plantation society had relied heavily on private tutors, Thomas Jefferson in 1779 introduced into the House of Burgesses a bill for public schooling under statewide legislation. He argued that the nature of democratic government in a republic demanded a common education for all. Every future citizen was a potential legislator and thus responsible for the common weal. As he wrote to John Adams in 1813: If his bill had been adopted—which it was not—the common elementary education it provided would have qualified the people "to select the veritable aristoi for the trusts of government . . ."[2] Such an education was to be seen as a legitimate, even necessary, task for a state or national community, and Jefferson wanted legislative action to bring this about.

In this effort Jefferson had not intended to deprive parents of their rights to determine the education of their children. As his friend, the

Frenchman Pierre Samuel du Pont de Nemours, stated in his plan for a national education, which he had written at the behest of and in cooperation with Jefferson: "It does not follow that the American Republic has assumed the power or the right to claim for the State, the Ruler, its delegates or anyone else in the world, the exclusive privilege of instruction." Du Pont de Nemours asked for respect for "the rights promised in the Constitution," and advised that private schoolmasters and parents alike should be allowed to teach children, provided that they use the state-prescribed text books and that their students pass the national examinations given to all children.[3] Jefferson's goal had been overall direction and supervision of schools and schoolmasters by a democratically elected government, not detailed control by an administrative bureaucracy.

After the end of the Revolutionary War and the establishment of the new republic, a number of American intellectuals took up Jefferson's cause of public education for all citizens of the new country. In the late 1780s and during the 1790s they composed treatises calling for a system of national education. Benjamin Rush of Pennsylvania, afraid that the heterogeneous character of the state's population would make for dissension and anarchy, called for "one general and uniform system of education." Pupils should be taught that they did not belong to themselves but were "public property." Rush also thought it possible that they could be converted "into republican machines." Thus he agreed with Noah Webster, his fellow countryman from Massachusetts, that "education should . . . be the first care of a legislature" and that in the American republic "knowledge should be universally diffused by means of public schools."

Others joined Rush and Webster and pursued the same theme. In Delaware Robert Coram complained that, except for New England, education was nowhere incorporated into the government. He demanded that government secure education to every child and that parents and guardians "be compelled to bind them [children] out to certain trades or professions . . ." Private schools should be turned into public ones, by which Coram meant schools supported financially by the public. Samuel Smith, who with Samuel Knox had won the prize offered in 1797 by the American Philosophical Society for the best essay "on a system of liberal education & literary instruction, adapted to the genius of the government . . . ," declared it to be "the duty of a nation to superintendent and even to coerce the education of children. . . . High considerations of expediency," he added, "not only justify but dictate the establishment of a system which shall place under a control, independent of and superior to parental authority, the education of children." Knox seconded Smith in calling for a uniform national system of education that included publicly supported common schools in each county, academies for secondary education, state colleges, and a national university. A state-appointed national board of education, Knox wrote, should assure the uniformity of the system.[4] It is clear that these educational reformers had little sympathy for the wishes of parents and would not have supported a call for school choice.

But just as with Jefferson's plan for Virginia, nothing came of these proposals. Local initiative had created the common schools, and until after the middle of the century they remained under local control. There is no evidence that the issue of national control over education was even discussed in contemporary periodical literature.[5] A predominantly rural population, proud of their recently acquired republican self-government, was determined to continue to resist outside taxation. Widely scattered, people lived in small towns, villages, and in the country. Their schools, too, were widely dispersed. This decentralization made local control all but inevitable.

Massachusetts recognized this fact in its education law of 1789, which recognized and sanctioned the district system. People were willing to pay local school taxes as well as parental rate bills. Their support led to a substantial increase in the length of schooltime throughout the year. Average pupil attendance and total enrollments likewise rose, especially for girls who were to be educated as the future mothers of republican citizens. Commercial cities favored common tax-supported over private schools, though that shift did not greatly affect the overall enrollment rates and was not characteristic for small towns.[6] By 1827 the General Court made local taxation for common schools mandatory, though in Boston primary schools had been brought under local public control already in 1818. Three years later the city opened the country's first public high school as a municipal institution.[7] It was not until the late 1830s and

1840s that school reformers set under way what has come to be known as "the public school revival." Then state governments began seriously to scrutinize and interfere with public education throughout the land.

But there had been a few states in which state government had begun early to play a role in school matters, New York being the most prominent among them. As in Massachusetts, in New York the state concerned itself first with preparatory and higher education. Pointing back to a 1774 Anglican-inspired draft charter of an American University in the Province of New York, an act by the New York state legislature in 1784 established a statewide administrative educational system called the Regents of the University of the State of New York. As revised three years later, the act asked the Regents "to visit and inspect all the colleges, academies, and schools" in the state and to incorporate newly founded colleges and academies. In 1813 the Regents were given authority to distribute the income of the State Literature Fund among the state's academies.[8] The Regents used that authority to favor academies that offered a classical education and mathematics and to discourage those that included utilitarian subjects like surveying, navigation, and bookkeeping. By their curricular preferences they sought to solidify a traditional system of social class, which was supported by classical preparatory and academic curricula that were, as they later admitted, "almost always the companions of leisure and wealth."[9]

With that policy the Regents had put themselves in opposition to the New York State Assembly. The assembly, representing the interests of parents who relied on the private academies for both an elementary and advanced English and modern scientific education for their sons and daughters, intervened in 1817. Its members rebuked the Regents who here played the role of an entrenched state bureaucracy representing the well-born and the wealthy. Assembly members, depriving the Regents of their right to charter academies and assuming that task for themselves, told the Regents that the academies were not to be viewed as preparatory institutions for liberal arts colleges, which concentrated on a literary and moral curriculum.[10] They were "people's colleges" that continued the work of the common schools on a higher level. The Regents, they said, should support studies "which fit the student for the useful occupations of ordinary life," especially for "that most important employment, the business of the school master."[11] In New York the State Assembly, representing the people, took up the cause of parents and rejected the policies pursued by the Regents.

In Georgia similar developments occurred. As in New York, a concern for the preparation of students for advanced education prodded the state legislature to subsidize private academies. A state board, called the Senatus Academicus and made up of the trustees of the University of Georgia and of state officials, was installed to determine the academies' curricula and to recommend their teachers and also to supervise the state's common schools. Just as in New York the Regents, so in Georgia the Senatus Academicus tried to promote the academies as classical schools. Again as in New York, the Georgia Senatus had to yield to the desire of the voters and was forced to support the academies as English schools and as finishing schools for Georgia's daughters. Civil society with its local concerns prevailed. In 1837 the legislature stopped the attempt to direct a statewide system when it decided to end the funding of the academies altogether.[12] At this stage the legislatures in New York and Georgia had made themselves mouthpieces for civil society and had limited the effectiveness of the statewide boards that had regulated secondary education.

Michigan was yet another state where the legislature ended the support of academies and turned its attention to statewide elementary education. In 1817, twenty years before Michigan was to be admitted to the union, a territorial legislature had set up a territorywide administrative educational system. The so-called Catholepistemiad was to supervise the territory's common schools, academies, colleges, and, eventually, a university. As in New York and Georgia, preparatory and higher education received primary attention. When the territory was admitted to the union as a state in 1837, nine academies received state funding. Though provisions were made for the funding of female, English, and teacher training departments in addition to the classical programs, most of the students attending the academies were boys who prepared themselves for entrance into the university. Just as in New York and in Georgia, popular dissatisfaction with the emphasis on university preparation and financial distress prompted the legislature to intervene. In 1846 it abolished state appropriations for the academies.

The timing of the legislative actions taken in New York, Georgia, and Michigan had been different. What had begun in New York State in 1795 with the assembly providing for state aid to common pay schools, had continued in 1804 with the creation of a permanent school fund, then reached its conclusion in 1812 with the creation of a state educational bureaucracy. This included the Office of the State Superintendent of Public Instruction, school commissioners and inspectors, and a system of school districts in the state's towns. In Michigan the withdrawal of state support for the academies in 1846 eased the way for these schools eventually to be replaced by public high schools. With that, State Superintendent Ira Mayhew argued, his office had taken over the administration of all of the state's activities in public common and secondary school education.[13] In Georgia a state public school system did not arrive until after the Civil War in 1872.

How had the actions in these three states replacing aid for the academies with the establishment of systems of public education and the creation of state superintendencies affected parents in their relations to the schools? As benign support of academies was replaced with increasing supervision and direction of the common schools, the parents of children attending academies had reason to feel slighted. But so had the parents of common school students whose room for direction of the school and their exercise of parental choice in common schooling began to shrink. New York's state school system with its divided responsibilities illustrates the point. While the state's school districts were charged with the maintenance of school buildings and the towns with the hiring and supervision of teachers, the state's responsibility extended to the distribution of school fund moneys to the towns.[14] Even so, parents were still required to support their children's attendance at common schools with rate bill payments. When then in 1814 the state assembly empowered the superintendent to add to his duties of distributing funds the power to make that distribution conditional on a town's raising an equal amount—an amount the town's voters could double, if they wished—it had given him the power to compel towns and districts to follow state mandates. Thus the assembly had effectively substituted state for local control. The sphere of parental influence and choice had been narrowed.[15]

The growing assertion of state legislative power over common education was to spread to other states and resulted in what has come to be known as the public school revival of the late 1830s and 1840s. Though there were exceptions—New York City is an example—it gradually brought to an end everywhere the prevailing practices of local and district control of common schools and thus diminished the influence taxpayers and parents could exert over their public schools. The revival began with the appointments in 1837 of Horace Mann as secretary of the Massachusetts State Board of Education and of Henry Barnard to the same position in Connecticut. The pioneers of the movement were motivated by what they saw as the excessive decentralization of the district system in rural areas. To them it was part and parcel of an unrestrained individualism that, in their views, threatened to undermine the newly acquired sense of national solidarity. Even in cities like Boston, as David Tyack writes of the 1840s, "the trustees of the primary schools were largely an independent, self-nominating, and self-perpetuating body . . ." They and sympathizers of the district system elsewhere, many of whom were members or supporters of the Democratic Party, fought Horace Mann and his supporters whose penchant for centralization they condemned as "associated . . . with King George, Prussian autocracy, and monopolies."[16] In 1840, however, they lost their battle when their bill for the abolition of the Massachusetts State Board of Education and the normal schools was decisively defeated.[17]

Horace Mann, Henry Barnard, and many Whig politicians felt that in rural areas the district system's individualism had led to unjustifiable differences in the support of common schools. In cities it had brought pauper legislation and private charity schools. Everywhere it showed a lack of concern for the common good. Arbitrary geographical circumstance and differences in parental income, religion, race, and social background favored or disadvantaged children. Only a public school system, so they argued, under the control of the best minds and warmest hearts of a community, could overcome these liabilities. Public schools would unify the people through a common education for all with a common language and a common religious faith that, to Horace Mann, would have to be a nonsectarian Protestant Christianity.

In order to accomplish this mission, common schooling would have to be compulsory for all. To leave school attendance to the discretion of the parents, stated a writer in *The Massachusetts Teacher* in 1851, would send children, and particularly the children of urban immigrants, not into our schools but "into our prisons, houses of correction, and almshouses."[18] Henry Barnard, finally, drew the final conclusion of this train of thought by making a person's right to vote dependent on one's having been schooled. "The right of suffrage," he wrote, "should be withheld from such as cannot give the lowest evidence of school attendance and proficiency."[19] As the 1840 vote in Massachusetts showed, these sentiments were not without effect. The rights of parents to freely determine the education of their children were to yield to the demands of the common good.

As an analysis of the 1840 vote shows, the debate over the kind of public school policy to be pursued in Massachusetts was linked to the voters' social class and economic status as represented in their political party affiliation. The Whig party, with 50.7 percent of its members being professionals, merchants, or manufacturers and only 30.6 percent being farmers, triumphed over the Democrats whose membership reversed these percentages. In the later party, 49.4 percent of the members were farmers and only 32.7 percent professionals, merchants, or manufacturers. Whigs, by and large, represented larger towns with high commercial activities, while Democrats were stronger in towns of less than 2,500 inhabitants in less-developed agricultural regions. To be sure, other considerations also entered into the debate. The Whig leaders' humanitarian motivations, the desire of workers for improved educational opportunities for their children, the belief of capitalist entrepreneurs that an extended public school system would insure economic and social stability, all played a role.[20] But social class as indicated by the voters' party affiliation, economic interests, and residence emerged as the most potent factor in determining the outcome of the vote in the legislature. The support of Whig professionals and manufacturers for the state board of education overcame the opposition centered on Democratic farmers in western Massachusetts. As Carl Kaestle and Maris Vinovskis have pointed out, the 1840 victory of the Massachusetts Whigs over the defenders of local control was to prove "prophetic of the direction public education took in the succeeding decades."[21] The centralizing tendencies of public school administration were to prevail.

Nonetheless, the victory of the centralizing forces was not complete. The conflict between the defenders of local and the advocates of state control continued. In the late 1850s and early 1860s it became evident again in the struggle over the opening and location of a high school in Beverly, Massachusetts. It shows that loyalty to district independence persisted among taxpayers and parents long after the state board had been firmly established. As Maris Vinovskis tells the story, "some Beverly citizens were unwilling to relinquish control of their local school districts to the town school committee, and many were even less willing to acknowledge the state's right to determine how education should be structured and run in their community."[22] Although these citizens and their like-minded successors never gave up their defense of localism, ultimately, they could not prevent the coming of state regulation of public schooling, the abolition of the district system, the introduction of normal schools, and the professionalization of teaching.[23]

Next to localism, loyalty to parochial schooling was the source of strong resistance to the advance of the public school movement. This was true above all in the large eastern cities of which New York may serve as the prime example. As pointed out earlier, the New York State Assembly had begun providing aid to the state's common schools in 1795. These appropriations continued for five years to local pay schools. While this policy worked well in rural areas, in New York City the large number of private pay school masters who could rightfully claim a share of the appropriations made a fair distribution almost impossible. So the city council decided to set aside five-sixth of the money for the construction of free public schools and to distribute the remaining one-sixth among private charity schools. But when the original appropriation's act expired in 1800 the assembly permitted the council to distribute all of the funds among the city's church-sponsored charity schools, leaving nothing with which to build new free public schools.[24] The council repeated this procedure in 1814 when it received the interest of the state's permanent school fund, which the assembly had created in 1804. Ten years later, however, the council ceased that distribution in response to the protest of the Free School Society. This group, sponsored by the Quakers, offered free education in its monitorial schools to the children of the poor and of free blacks. It argued that it held a superior claim to the state money because, by serving children without regard to their religious affiliation, it had taken on a public responsibility. The council agreed, and in 1826 recognized the Free School Society as the Public School Society.

This shift in policy aroused the ill will of New York's Catholics who were afraid that their children's faith was endangered in the schools of the Public School Scociety. When word spread that this organization planned to segregate Catholic children in their schools, Catholics were alarmed that the Society intended to launch a missionary campaign and attempted to convert the pupils to Protestantism. Catholic fears were heightened during the 1830s by a rising wave of anti-Catholicism and nativism that was accompanied by inflammatory publications, burnt convents, and street riots in several cities. Bishop Hughes of New York warned the Common Council that "if a single Catholic Church were burned in New York, the city would become a Moscow," alluding to that city's conflagration of 1812. Governor William H. Seward repeatedly spoke out in support of New York's Catholics and recommended the establishment of confessional public schools where immigrant children could be taught by teachers of and in their own language and faith. He did so, he said, "less from sympathy, than because the welfare of the state demands it, and cannot dispense with it." But to no avail. New York's aldermen rejected a Catholic request for public funds for their schools on the grounds "that taxation of all sects for the benefit of one is a violation of the rights of conscience."[25]

The issue finally found a solution of sorts in the state legislature in Albany. In 1842 the assembly passed the Maclay bill which decreed that in New York City, though not outside its limits, school policy was to become a matter of community control. The communities were to be the city wards whose citizens were to elect representatives to the boards of school districts, which were to be coterminous with the wards. A citywide Board of Commissioners of Common Schools, to be made up of elected representatives from each ward, was to supervise the city common schools, which were to include those of the Public School Society. Thus while parents and citizens had gained a direct voice in governing common schools in the wards or districts, theirs was a limited kind of "local control" supervised and checked by a central Board of Education. It was furthermore held in check by the Maclay bill's prohibition against sectarian teaching in the public schools. But parents who desired their children to receive such instruction did have a choice. They could send their children to one of the Protestant denominational charity schools, to private proprietary pay schools, and eventually also to Catholic parochial schools.[26]

While the Public School Society had now come under the control of the democratically elected Commissioners of Common Schools and thus had lost its monopoly, so strongly resented by Bishop Hughes, the new arrangement did not satisfy the bishop. He had not received public funds for Catholic schools, and Catholic students in the city's schools were still exposed to a Protestant kind of nondenominationalism and to the reading of a Protestant Bible. This did not change either eleven years later in 1853 when the Public School Society officially closed its operations and handed its schools over to the city's public school system. Now, the bishop complained, Protestant nondenominationalism more often than not had given way to "godlessness." He therefore embarked on a vigorous campaign to build up for his parishioners an independent Catholic parochial school system. To his thinking, "in this age and country, the school is before the church." His initiative came to full fruition in 1884 when the Third Plenary Council required of all bishops and priests to build parochial schools and of parishioners to enroll their children in them.[27] By 1890 that system enrolled 626,496 students and was, as David Tyack writes, "the largest 'alternative school system' in the United States."[28]

With the passing of the Maclay bill in 1842 New York City had arrived at a system of school choice that, though not entirely free in all its aspects, could serve as a model for the country's larger cities. New Yorkers could now send their children to Protestant denominational charity schools, to private proprietary pay schools, to the city's common district schools, including those of the Public School Society, and to the now beginning Catholic parochial schools. To be sure, there were limits to the freedom of this choice. Financially free schools were available only to poor parents who sent their children to the Protestant charity schools and to all parents who made use of the new public school system. Private pay and, in most cases, Catholic parochial schools required tuition and fees. Moreover, following the meeting of the Third Plenary Council in 1884, choice was no longer available for Catholic parents who desired to obey the council's decrees. However, in all these arrangements school control by local communities was still the rule. In New York City's district system these communities were determined geographically; in private arrangements they were set denominationally or financially. Both these community boards and the central city Board of Education served as agencies of civil society. They, not the state superintendent, effectively regulated New York City's schools. In New York city's community control system the public school revival had found an effective roadblock and parents had been given a limited opportunity to exercise school choice.

When we now take a look at the formation of educational policy in Prussia we shall note the contrast between popular involvement in school affairs in the American colonies and the United States on the one hand, and the role of Prussia's government and of its intellectuals and administrators on the other. Royal decrees and academic debates played a far more central role in Prussia than they did in America. But administrative directives also met determined resistance from local school authorities on whose shoulders rested the burden of financing and maintaining their schools and on whom the state bureaucracy had to rely for the execution of its wishes on both provincial and local levels. Representatives of churches, municipalities, boards, and societies as well as private patrons remained a potent counterforce to state control and direction and frequently resisted or ignored state directives. In this hybrid system of school governance parental school choice, if it could assert itself at all, existed only where local school sponsors were able to provide and permit it.

Prussia's official school policy began in 1717 when King Frederick William I attempted to introduce compulsory schooling in his realm. In subsequent decrees of 1736 and 1763 Prussia's government urged patrons of the nobility to see to it that children in their villages learned to read and write and became acquainted with the rudiments of Christian doctrine.[29] Prussia's General Land Law of 1794 declared Prussian schools and universities to be "institutions [*Veranstaltungen*] of the state that intend to instruct youth in useful knowledge and sciences" and which should be established only "with prior knowledge and permission of the state." It further decreed that "all public schools and educational institutions are subject to the supervision of the state and to its examinations and visits at all times." Parents unable to provide for the necessary instruction of their children at home were required to send them to school after they had reached the age of five. Local patrons and secular and ecclesiastic village authorities were to grant exceptions for those children who, because of domestic business, could not attend the ordinary school hours during seasons that were given over to certain necessary labors. For them alternative hours were to be arranged on Sundays and other suitable times.[30] These exceptions referred to both agricultural and, until the 1840s, to industrial child labor as well.

By the end of the eighteenth century Prussia's government had set up a general framework for instituting compulsory education, but had left the execution of the policy in the hands of local authorities and private school sponsors. After Prussia's disastrous defeat by the French army in 1806, reformers enacted legislation intended to enforce supervision of all schools by state administrative agencies. Through the newly created Section for Culture and Instruction they set up in 1810 and 1811 a system of school deputations in the provincial districts and in municipalities and rural communities. These agencies were to decide the types of schools to exist, when and where they could be opened, and who was to run them. They were to authorize curricula and methods of teaching, and to regulate examinations and access to higher schools and universities.[31] The determined resistance of the traditional patrons of the elementary schools, however, forced the deputations to compromise between asserting the section's directives and protecting the sensitivities of local school authorities. For the supervision of the higher schools the section vested existing church consistories with the authority of provincial school commissions. The result was that compulsory schooling under state supervision was neither universal nor compelling.

The lax enforcement of "compulsory education" (*Schulpflicht*) was due not only to the opposing pulls of central directives and local interests on the provincial and municipal school deputations. We can attribute it also to the exceptions permitted by the law itself and, what was undoubtedly the strongest factor, to the expectation of the Prussian government that elementary schools—which, according to the General Land Law were to be tuition free—be financed in the main by local associations, school societies, and municipalities. Local regulations, however, often ignored the Land Law and differed from each other and from province to province. Thus it is not surprising that compulsory attendance at elementary schools remained problematic, and parents felt no need to worry much about school choice.

The phrase of the General Land Law of 1794 that declared schools to be "state institutions" remained similarly ineffective. Schools in cities were expected to be and were in fact administered and financially supported by religious congregations or the municipalities in which they existed. In the countryside villages, school societies, and private school patrons functioned as school sponsors and supporters while village priests or parsons served as supervisors. Private schools, though nominally under state supervision, continued to exist as property of their owners.[32] That despite the legal framework of state direction, schools were de facto in the hands of the agencies of civil society, was underlined by one of the authors of the Prussian General Land Law who maintained in a public address that, though they were instituted by the state, schools should not be seen as state institutions.[33]

As a result of the state's unwillingness to devote more than a bare minimum of its financial resources to the upkeep of the country's schools, these were of many kinds, and the two-stage organization of general education in elementary schools and *Gymnasia* projected after 1810 by the Section for Culture and Instruction existed as an idealized prescription only.[34] In rural areas, where the majority of the country's population lived, local clerics and landed proprietors maintained their traditional influence until late into the nineteenth century and offered no opportunities for parents to choose their children's schooling. In towns and cities a child's school attendance was largely determined by the parents' social class. For the children of day laborers and the poor the choice was restricted to charity and publicly supported pauper schools, factory and Sunday schools, or to go without schooling altogether. Parents of the lower bourgeoisie could turn to *Bürgerschulen, Mittelschulen*, and *Realschulen*. These schools served either as advanced elementary schools of general education or they provided for the needs of young people who sought a practical-oriented, vocation-preparatory type of schooling. Parents could choose as their finances would allow. The pride of the community and the choice of the city's business and professional groups were the higher schools, *Latein-* and *Gelehrtenschulen* as well as *Progymnasia* and *Gymnasia*.[35] Here parental influence could assert itself and help determine the school a child would attend. All of them prepared their students for careers in the middle levels of state service, and the *Gymnasia* entitled them to a one year term of military service instead of the usual three years. Those students who had passed the *Gymnasium's* leaving examination thereby obtained the right to attend a university and thus to gain access to the professions and the higher levels of state service.

Toward the end of the eighteenth century the differences set by locality, economic standing, and social class prompted several Prussian statesmen, administrators, and educators to address the public on the country's educational efforts. As disciples of the Enlightenment many of these men espoused rationalist and progress-conscious ideologies. Their reflections were, in many ways, quite similar to the treatises of Benjamin Rush, Noah Webster, Robert Coram, Samuel Smith, and Samuel Knox in the United States. Motivated by concerns over the country's lack of economic development and internal cohesion these men deplored existing efforts as slow-moving, inefficient, frequently corrupt and pointing in too many different directions. While few of them wanted to abolish the sponsors' control over their schools, they expressed impatience with what they saw as a waste of resources.

Other critics were warmhearted humanists who wanted to alleviate misery and poverty by bringing schooling to all in a fair and equitable manner. They agreed that this could be done best by centralized administration and supervision of public schooling. The rise of nationalism at the century's turn only added impetus to their endeavors.

This push for administrative reforms began with teachers of the cameralist and police sciences like Johann Heinrich Gottlob von Justi and Georg Heinrich Zincke. These men had recognized the political connection between schools and a country's constitution already in the first half of the eighteenth century.[36] When their student Johann Heinrich Bergius wrote in 1774 that the state could no longer rely on parents to educate their children toward love of country, industriousness, and productive competence, he implicitly rejected any notion of parents having a voice in the determination of school policy. He saw central direction as the answer to uncoordinated local efforts. The "police," he wrote, that is, public authority, had to institute and supervise the schools.[37]

The decisive step in translating academic treatises and recommendations into effective administrative regulations was taken by Karl Abraham Freiherr von Zedlitz, a onetime student of Georg Heinrich Zincke. Zedlitz was Prussia's minister of justice responsible for the country's religious and educational affairs administered by the Lutheran Church. He believed his office as it functioned to be unworkable because it gave him little information on secular educational developments and suffered from the absence of an institutional memory. In 1787 he published his *Suggestions for Improving Education in Kingdoms*. In it he stressed the need for increasing the number of public schools and placing them under central secular supervision.[38] Arguing that the purpose of schooling was to prepare everyone for life in society Zedlitz, a disciple of Enlightenment educational thinking, advocated improved opportunities for schooling within the existing class system of Prussian society.

Zedlitz recognized three classes of society and argued that each required a corresponding type of school. The children of peasants were to be sent to country schools; those of urban wage earners to *Bürgerschulen*, and those of professionals and noblemen to higher or

Gelehrtenschulen. In addition, there were to be institutions for the training of teachers. By associating professionals with noblemen, Zedlitz acknowledged the emergence of an upper educated bourgeoisie (*Bildungsbürgertum*) whose children sought advanced education beyond what they and the children of the wage-earning and entrepreneurial lower bourgeoisie (*Besitzbürgertum*) received in the *Bürgerschulen*.[39] To put the educational system in effect, supervise and inspect its institutions, including the existing private ones, and to examine its teacher candidates, Zedlitz recommended the creation of an *Ober-Schulkollegium*, a state higher school board. It was to consist of five appointed educators, directly responsible to the crown. The king signed off on Zedlitz's proposal in the same year and the *Ober-Schulkollegium* began its work. Its impetus was never to be lost and its basic thrust of central secular direction of the nation's schools has been felt to this day.[40]

The issue of state-sponsored education, however, was contentious from the beginning and has remained so for two centuries. Traditional school sponsors, churches, localities, and private patrons resisted what they saw as the "medling" of the newly established *Ober-Schulkollegium*. Its decree of 1788, regulating access to the universities by requiring a high school leaving examination, displeased the directors of existing Latin schools who feared—needlessly, as it turned out—for the continuing existence of their institutions.[41] Debates and discussions ensued among many of Prussia's intellectuals, some of whom remained ambivalent about the social implications of state activity in education and others who expressed their opposition.

Peter Villaume, 1788 *Gymnasium* teacher and minister of the reformed French congregation in Berlin, was ambivalent. His contradictory views for and against centralized, monopolistic state education have a contemporary ring. Would not state interference in and supervision of education make it the state's business to assign some children to a privileged education and condemn others to continue living in poverty, he asked? Would this not violate parental responsibilities? And should not educational questions be left to the expertise and discretion of educators? But he soon found other arguments to overcome his scruples. Differences in class structure were inevitable, he reasoned. Therefore the state should educate everyone according to his or her estate. Only in this way could the state decrease the unhappiness of individuals who were unprepared to accept and live within their conditions. To the rights of parents Villaume juxtaposed the rights of children which it was the state's duty to protect. And as for the expertise and discretion of educators, they were to be assured and demonstrated by state-prescribed examinations and supervision. The new leaving examinations would accomplish this for the higher schools. After all was said and done, Villaume concluded, it was the state's duty to concern itself with the education of its children.[42]

In the same year in which Villaume penned his remarks, Johann Christoph von Wöllner, Zedlitz's successor as official in charge of Prussia's religious and educational affairs, published his *Religionsedikt*, which imposed a rigid censorship on Prussia's teachers and university professors. Wöllner's rise to power signaled a trend to conservatism and religious orthodoxy in Prussia's policies. Among the country's liberal intellectuals Wöllner's edict stirred opposition and, four years later, led Ernst Christian Trapp and Wilhelm von Humboldt in separately penned essays to plead for limits on the state's power to regulate schools. Trapp, a leading philanthropist philosopher and, in 1779 in Halle, the first professor of education at a German university, wrote that there should not be any state-directed schools at all. The state, he explained, should support but not supervise public schools which were to be open to everyone. As a spokesman for what today we call a free market in education, Trapp asked for state permission of private schools and instruction and for the elimination of any state-sanctioned supervision of religious schools. Furthermore, the state should not be permitted to determine the schools' curricula and the education of teachers.[43]

In basic agreement with Trapp's views was Wilhelm von Humboldt's famed essay, "Ideas for an Attempt to Determine the Limits of the State's Effectiveness." While the complete essay was not published until 1852, Chapter VI, which included Humboldt's thoughts on national education, appeared in the fall 1792 issue of the *Berlinische Monatsschrift*. In it Humboldt rejected the argument that examples from republican antiquity could serve as models for monarchical Prussia. In a republic, he wrote, citizens were in charge of their own affairs and could well be entrusted with governmental authority over themselves. But in a monarchy they lived under the thumb of royal authority and needed to be safeguarded against arbitrary power. The people of Prussia needed an education that would raise the potential of individuals to higher levels. This could not be achieved in schools under state control. State-directed education, Humboldt wrote, sacrificed the individual to the citizen. It thought of citizens and subjects, not, as did private educators, of human beings. It produced a uniform character type and led to imbalance in the body politic. Thus, concluded Humboldt, public education lay outside the limits to which the state should be restricted.[44]

Seventeen years later, after the country's defeat by Napoleon, Humboldt's actions seemed to belie the convictions he had expressed in 1792. The king then asked him to take charge of the reconstruction of Prussia's schools and universities and, despite his earlier unwillingness to place the administration of a country's educational affairs into the hands of the state, Humboldt accepted the offer. He asked himself whether, under the circumstances then prevailing in defeated Prussia, state action might not be necessary to initiate the country's educational reconstruction. Though he was afraid that, if he accepted the call, he would only sacrifice himself and his views without being able to achieve anything useful, he also feared that, if he refused to serve, he would be accused of being ungrateful, of lacking love of country, and of deserting those in need.[45] So he consented to serve, persuading himself that through a temporary reliance on state action he might in the end bring to fruition the educational reforms he had long desired.[46]

Throughout his term of service Humboldt held to his aversion toward state action and labored to strengthen the agencies of civil society. As head of the Section on Culture and Instruction in Prussia's central administration he expressed that attitude in his letter of March 14, 1809, to C. Ludwig Natorp, his deputy in charge of the reform of teacher education:

> You will share my opinion, [he wrote,] that very much can be done by local communities and, when possible, by the people . . . It is wrong to demand that the state do everything; it is salutary for the

> nation's independence when great, beneficent institutions grow, as it were, by themselves from the nation's lap.[47]

Similarly, in his 1809 plans for the rebuilding of schools in Königsberg and Lithuania he placed the emphasis on school policies designed for city and province. In his 1810 report on financing Prussia's schools he asked that a separate national school fund be administered by local authorities and kept separate from the state treasury.[48]

The realities of Prussia's reconstruction, however, forced modifications in his plans. They ruled out his proposal for a locally administered national school fund. Like Villaume before him, Humboldt found himself in a position where he argued the case against and for state direction of schooling. Though his heart remained with his earlier views, existing conditions and the requirements of his office compelled him to use the instrumentalities of the state for the realization of his ideals.[49] He resigned his position in June 1810, sixteen months after he had been appointed.

The most uncompromising advocate of state direction and supervision of education among Prussia's intellectuals was the philosopher Johann Gottlieb Fichte. Relying on the statement in Prussia's General Land Law of 1794 that schools and universities were instituted by the state and interpreting that phrase to mean that they were state institutions, Fichte argued that a system of national education was necessary to safeguard the stability of society and to enable the state to shape a national consciousness among its citizens. In his *Addresses to the German Nation*, a series of public lectures he delivered in 1807 at the Prussian Academy of Sciences in Berlin, Fichte sketched a vision for a unified state educational system in which the university appeared as the capstone of the whole.[50] He appealed to his listeners and readers to rise above their chagrin over the defeat of Prussia and view the catastrophe as an opportunity to begin building a German nation. For such nation building, a state-directed public education system would educate the German-speaking peoples to national consciousness, liberation, and service. A common philosophy, rationally conceived and uniformly embraced and adopted by all, would lead the people to reject their private and parochial interests. National education in state-directed schools was to shape the German nation in the nineteenth century.[51] In an educational system ruled by such an approach there could scarcely have been room for parental school choice.

At the height of the liberal reforms in the years from 1810 to 1819, Trapp and Humboldt were not alone in their attempts to preserve the authority of local institutions of civil society and to protect the interests of children and parents against state power. The Königsberg pedagogue Johann Friedrich Herbart wrote in 1810 that the state, being a collection of many diverse interests and elements, could only produce public schools that reflected this diversity and thus created divisions and separations among its children. Of necessity such schools dealt with masses of children and thus resembled factories. At best they were makeshift devices whose effectiveness he doubted. Herbart argued instead that education, concerned with individuals and their development, should be offered by teachers working as tutors in homes and small public, but not state-directed, community institutes. In these institutes which, in their social makeup would reflect the diversity of the community, parents would wield a large amount of discretion in choosing their children's tutors. They would thus come as close to enjoying the benefits of school choice as was possible under nineteenth-century circumstances. Herbart's plan, however, was never realized in practice.[52]

Like Herbart, the Protestant theologian, philosopher, and university reformer Friedrich Schleiermacher believed public education should lead to vibrant local communities and community consciousness. If, he wrote in 1814, state participation in national education promised to achieve this result, he would favor it. But whenever this was unlikely to be the case, the state should hand education back to the people. However, such return of education to civil society, Schleiermacher warned, must make sure that for boys education remain public and not be handed over to private promoters. Private education, he wrote, created nothing but arbitrariness and was derived from a longing for capriciousness and a lack of community feeling. It could be appropriate for girls whose lives were to be lived in the private, domestic sphere. But boys, the country's future civil servants, were to become leaders in an active civil society-which, cooperating with the church and with science, worked indirectly also with the government of the state. For boys, therefore, national education for public purposes was to take place as a vibrant part of civil society, not of government.[53] While one may assume that in Schleiermacher's view parents as members of civil society had at least a limited voice in determining the public schooling of their sons, they were left to their own devices to arrange for the private schooling of their daughters.

The reality of Prussia's educational policy developed along lines different from those suggested by Humboldt, Herbart, or Schleiermacher. It was characterized by a vigorous interplay between state and civil society. Beginning in 1787 with the creation of Zedlitz's National Higher School Board, continuing with the administrative incorporation of the country's schools in the Ministry of the Interior in 1808, and leading to the activities of a Cultural and Educational Ministry in the years from 1817 to 1825, the Prussian state claimed increased supervisory and directive activities over its schools. Still, this development was kept in check by the continuing presence of defenders of local school authorities and, more importantly, by the provision of the General Land Law to finance schools by local rather than state taxes. This left decisions concerning the number, construction, and upkeep of local schools, the pay of teachers, the average size of classes, and the enforcement of compulsory education in the hands of local school committees or societies.[54] The interplay between state and society regarding the maintenance of schools continued into the second half of the century. An organized teacher movement, dissenting theologians participating in the administration of schools, and the members of urban school commissions who were usually appointed by the town's mayor but needed confirmation by provincial or state authority, all were players in this game.[55]

However tense relationships may have been between advocates of state or civil society, there was general agreement that the agencies of state and civil society, including those of the established Lutheran Church, were public in nature and were designed to serve public purposes. Schools served the public, existing schools in private ownership notwithstanding. Parental concerns, essentially private in nature, thus were of secondary importance, even to local school authorities. They were even less a concern to the members of Prussia's educational administrative bureaucracy—the German *Beamten* (civil servants).

The *Beamten*, whether liberal or conservative in their social and political views, saw themselves as spokesmen for the enlightened interests of the whole represented in the state. They interpreted the Prussian Land Law to mean that it was the state's duty to see to it that all children were educated in public schools for their roles as citizens. They argued that it was their responsibility, in cases of parental neglect, to represent and assert the educational interests of children. They differed among themselves in details. As pointed out before, the followers of Humboldt wanted a two-part system of general education for everyone in elementary schools and a continuation of advanced general education in higher schools for those who could muster the necessary financial and intellectual resources. Others promoted various middle school types for a more vocational-preparatory education. As a group, however, the *Beamten* constituted the strongest force for gradually asserting and solidifying the power and control of the state over Prussia's educational system.

This also meant that despite the many eloquent pleas penned by the defenders and proponents of school choice among Prussia's intellectual and administrative elite, the *Beamten* generally held that school choice was not an option that could be introduced at the state level. Local school providers, however, could, and in many cities, did offer school choice whenever their financial conditions permitted it and taxpayers, civic groups, churches, and private school sponsors offered the opportunities. For parents limits thus were set not so much by public decree as by their class and economic standing in their community. Limited as it may have been by state law and local financial and entrepreneurial resources, school choice did exist.

Chapter 2

The Systematization of Public Education

The establishment in 1812 in New York State of a State Superintendent of Public Instruction and the appointments in 1837 of Horace Mann in Massachusetts and of Henry Barnard in Connecticut as secretaries of their respective state boards of education had laid the groundwork for what was to become the American system of public education. The New England example found imitators in other states, and by mid-century the so-called public school revival had begun to extend across the country. Only the antebellum South lagged. There it took hold during and after the 1870s. Initially the powers of state superintendents were limited to gathering and distributing information and to encouraging and persuading local and district school boards to strengthen the common schools. As time went on, however, the state superintendents' administrative powers grew and eventually led to a nationwide decline in the control of local school boards over their district schools. For parents this meant that the further removed from their residence the decisions concerning public schooling were made and the more they were predetermined by state regulations, the less effectively could they assert their choice and preferences over their children's education.

In the United States as in Prussia this centralization of school administration brought with it a progressive systematization of schooling that proceeded over most of the nineteenth century. But whereas in Prussia municipal initiatives focusing on trade and vocational education often had moved at cross-purposes with the more academic and humanistically oriented school plans favored by the state education ministry, in the United States city and state superintendents who often were former politicians or Protestant ministers did not have to cope with opposition from humanistically trained civil servants. Motivated by a strong commitment to Christian ethics and service to their fellow countrymen and-women, they did not usually look upon their appointments or elections as calls to lifelong administrative careers. They lacked the *esprit de corps* and sense of duty a Prussian *gymnasium* and university education had instilled in their German counterparts. They were less inclined than their German colleagues to impress their views on local school board members, councilmen, and parents. They preferred instead to work with the local public school boosters among the cities' petite bourgeoisie parents and to seek compromise and adjustment to local conditions.

By the middle of the nineteenth century the schoolmen in American cities, whether they were principal teachers or the local or state superintendents, came to rely heavily on cooperation with the members of the leading local elite of small businessmen, newspaper editors, and professionals who still held fast to the concept of local control of neighborhood schools.[1] They jointly sought to cope with the continuing progress of urbanization and industrialization, which called for providing new educational opportunities and changed the character of many urban neighborhoods. In Boston, for example, responding to the pressure of middle-class parents, the city added a high school for boys in 1821 and one for girls in 1826 to the existing Latin and reading and writing schools. By mid-century the arrival of immigrants changed the picture again. The Boston School Committee announced in 1847 that the children of these great masses of foreigners "who are not educated, except to vice and crime," had to be "made inmates of our schools . . ." The city's established businessmen and professionals argued that it was more economical to educate these children in the city's schools than to support them in its prisons.[2] While the changes became visible in local settings they represented national trends that demanded national adjustments.

In the countryside systematization gradually moved the control over schooling from civil society to the states and, to some extent, to the federal government. After mid-century its greatest forward push came during and after the Civil War with the freeing of the slaves and, in 1867, with the creation of the U.S. Bureau of Education. Under its first commissioner, Henry Barnard, an inspired promoter of state influence in public education, the bureau laid the institutional foundation for what eventually became federal oversight of schooling. In the same year, congressional radical Republicans sought to require the ex-Confederate states to open their schools to children of both races. While these efforts helped to make public education a reality in the Southern states, they could not assure equal treatment for black children. Whites, dominating local school boards, saw to it that black children, if they attended schools at all, were segregated from white children in underfinanced schools of their own. In 1896 the U.S. Supreme Court in *Plessy v. Ferguson* made this kind of segregation the law of the land under the specious doctrine of "separate but equal," where segregation was real, but equality was not. School choice across the racial divide was an option for neither black nor white parents.

During the reconstruction period following the Civil War the partisans of state and national direction of public education became persuaded that theirs was the preordained course the nation's educational policies were to follow. The Freedman's Bureau, established by Congress in 1865, opened over 4,000 schools for the freed former slaves and supported the educational efforts of private missionary and freedmen's aid societies. This massive effort by missionary groups and the bureau to bring education to the ex-slaves rested on the assumption that the freedmen, untutored and uneducated in the ways of white society but eager to learn, would need start-up help from white teachers sympathetic to the freedmen's cause. As Louis Harlan put it, the public school therefore was "to stand *in loco parentis* for the freed Negro . . ."[3] But it would come to be a question whether missionary teachers and teachers of the Freedman's Bureau or the ex-slave-holding white members of local school boards would represent the public school Harlan referred to.

But as the American Missionary Association announced in 1865, its agenda comprised more than the effort to open public schools for the freedmen. The association was also intent upon creating a truly national system of public education. As one historian noted: "The elimination of slavery and the subjugation of the rebels provided reformers with a grand new opportunity to influence the destiny of the entire country."[4] This combination of motives—aid for the freedmen and creation of a national system of public education—did not detract from the high-minded devotion of the teachers and aid workers to the cause of the ex-slave. It did, however, raise the question whether, once the initial period of need had passed, the American Missionary Association should continue to promote the public school in its parental role or whether it should seek to transfer responsibility to black educators and parents. Both Louis Harlan and Carter Godwin Woodson had wondered whether help, not offered as an inducement to self-help, was of lasting effectiveness, particularly when, as Woodson pointed out, "it was the liberated Negroes themselves who, during the Reconstruction, gave the Southern States their first effective system of free public schools."[5] W. E. B. DuBois stated unequivocally that "public education for all at public expense, was, in the South, a Negro idea." He could point to Georgia where, by 1867, 191 day schools and 45 night schools had been established and were in operation by Georgia blacks.[6] Similarly, Mississippi blacks had opened schools of their own even before the war had ended and Northern teachers had arrived. Many of them refused to accept Northern assistance and maintained their independent existence as competing educational alternatives to the Freedmen's Bureau schools.[7] Their persistence was justified when in subsequent decades the courts upheld segregation enforced by local school boards and state educational authorities and black parents enjoyed neither equality of educational opportunities nor choice in the selection of schools for their children.

One may also wonder to what extent congressional legislation of the 1870s and 1880s, intended to aid the Southern states financially in order to combat illiteracy, was prompted by reasons that had little to do with aiding the former slaves. After the divisive and fratricidal rupture of the Civil War, concern for national unity played an important role in the thinking of many Americans. Reliance on sectarian schools, they thought, would undermine national unity and preserve racial segregation. When President Grant in 1870 urged the

Congress "to take all the means within their constitutional powers to promote and encourage popular education throughout the country," he had in mind more than the establishment of public education systems in the Southern states, systems that would include the newly freed slaves.[8] Representative George F. Hoar of Massachusetts followed up Grant's initiative with a bill to establish a "national system of education," which, if a state was negligent in carrying out its provisions, would have given the president the power to appoint a federal school superintendent who was to supervise the state's efforts. Fellow Massachusetts legislator, Senator Henry Wilson, praised Hoar's bill as a means of unifying and educating the American people. No state, he wrote, "should have the power to prevent the national prerogative from being exerted in that direction." And, picking up the theme of Prussian superiority in public schooling, a theme that had been sounded thirty years earlier in the battle over the proposed abolition of the Massachusetts Board of Education, he credited Prussia's system of compulsory education with being responsible for the country's 1871 victory over "ignorant, priest-ridden and emasculated France. . . . The lesson should not be lost on the American people," he added, "especially upon the Republican party."[9] Though the legislation failed, its proponents left no doubt that they were determined to shift the controlling powers over America's public schools from local and district officials to professionals in state and, eventually, national offices.

Hoar's bill was prompted by concern over national unity and was based on the assumption that schools played a role in raising national consciousness and had therefore to come under federal oversight. But American Catholics viewed it as intended to rule out any chance of public support for their parochial schools. This was a matter that until then had been argued and adjudicated on state or local levels. Now, Catholics feared, such arrangements were meant to give way to "a system of universal and uniform compulsory education" for all American youth.[10] Protestants, too, voiced such fears and opposed the bill because it threatened the defenders of states' rights as much, if not perhaps more, as the advocates of nonpublic schooling.

But though the bill was defeated, efforts continued to discourage the growth of nonpublic schooling, especially that of Catholic parochial schools. Late in 1875, Representative James G. Blaine of Maine offered a constitutional amendment that would have made it illegal for any public funds raised or intended for the use of public schools to come under the control of or for the use of religious sects or denominations. When Blaine's efforts failed, further attempts to prohibit the use of public funds for schools under sectarian control were made by Republicans in their 1876 platform and, during the 1880s, by Senator Henry William Blair of New Hampshire. All of these efforts likewise failed, not because of resistance to the bitter anti-Catholicism expressed by its proponents, but because of constitutional scruples, fiscal considerations, and sectional rivalries.

It was not until 1925 that the U.S. Supreme Court in *Pierce v. Society of Sisters* declared unconstitutional a national policy to make all schooling an exclusive responsibility of the state. The Court affirmed that nonpublic schools had a constitutional right to exist and parents had the right to enroll their children in them.[11] Without using the term, the Court had endorsed school choice. On the related question, whether public funds could be used for and by non-public schools, the Supreme Court ruled in 1947 in *Everson v. Board of Education of Ewing Township* that "no tax in any amount, large or small, can be levied to support any religious activities or institutions, whatever they may be called, or whatever form they may adopt to teach or practice religion."[12] The *Everson* case thus leant support to the position of the anti-Catholic campaigners of the 1870s and 1880s who had charged that by appropriating public moneys to sectarian schools, states had ignored the Fourteenth Amendment's extension of the First Amendment's religious establishment clause. If school choice was to mean that parents could freely and without additional expense to them enroll their children in schools that taught "tenets and faith of any church," the *Everson* case had declared such school attendance unconstitutional.

Though the post–Civil War debates over nonpublic education and the degree of state influence over schooling were prompted by developments in the South, they affected all parts of the nation. In Wisconsin, to take one example, common schooling had been locally financed during its territorial period through a mixture of rate bills, tuition fees, and town or county taxation. At the state constitutional conventions in 1846 and 1848 the delegates opted for public district schools that were free of fees and bills to the residents of the state. The delegates refused to permit monitorial schools with regimented classrooms, and they would not tolerate charity schools, which segregated the children of the poor. They likewise spoke out against sectarian instruction that might offend religious sensibilities. The schools were to be open to all state residents between the ages of four and twenty, regardless of race and gender. For financing the schools the delegates preferred to rely as much as possible on local funding because they were afraid that contributions from outside would undercut the residents' interest in their schools. But they also were loath to forego the support they could expect from the state school fund that had been accumulating from the receipts of federal land sales and from other federal grants. So they decided to set township and city taxes at half the amount a town or city would receive from the state school fund, and to ask school districts to tax themselves for the construction and maintenance of schoolhouses. To the state school fund they assigned responsibility for the financing of teacher education in county academies, public normal schools, and a university.

In the assignment of offices the delegates demonstrated their preference for persons known in their communities and in the state. They therefore assigned the organization of school districts and the examination and certification of teachers for the district schools to township superintendents who, it was assumed, as elected town officials were personally familiar with the town's inhabitants and could best represent their interests. For similar reasons they preferred an elected over an appointed state superintendent. Election presupposed the candidate resided in the state and was familiar with state conditions. Appointment, on the other hand, would make it possible to bring to Wisconsin a candidate from abroad. The argument for election won handily.

While the concerns for local control of schooling and lay input into educational policies were obvious in the deliberations of the constitutional convention, issues that required state attention or arose from the interests of educators soon came before the legislature. The State Assembly opened a State School for the Blind in 1850 and, ten years later, a State Reform School for Boys. In 1857, in accordance with the Constitution, it appropriated funds for teacher education to colleges, academies, and normal schools. Two years later it funded teacher summer institutes, which had been inaugurated earlier by the state superintendent. With the superintendent's encouragement, teachers founded the State Teachers' Association in 1853 and issued the first number of the *Wisconsin Journal of Education* in 1855. The statewide professionalization of teaching had begun.[13]

The push and pull between local and statewide interests in education manifested itself most clearly in the battles fought over the introduction of the high school. Support came mainly from cities and villages where parents saw in the high school a means for their children to advance socially and economically at taxpayers' expense, and where businessmen believed that such a school would boost the commercial welfare of their community.

A few parents appreciated the opportunity to have their children prepared for college without having to send them away to a private academy or boarding school. Public school educators were in favor because, if nothing else, an expanding school system statewide augmented their professional responsibilities and prestige. While they spoke of the "connecting link" the school could provide between common school and university, they were well aware that relatively few students planned to attend for that purpose. More Wisconsin youngsters wanted to be "prepared for life" than for college, and the high school stood to benefit more from youngsters who harbored business rather than academic ambitions. Add to this that for girls the school's "preparation for life" included in large measure training for a position as elementary school teacher, and it becomes understandable why schoolmen organized so many early high school classes as normal classes. They came to rely on them as their very own source of substitute and future common school teachers.[14]

Opposition to the high school came for the most part from childless taxpayers and from parents in the rural townships. On the farms widely dispersed over the state parents were not easily persuaded that a high school could benefit them or their children. Farming, they believed, was not learned in a school. Few believed that their children needed or wanted a college or university education. Besides, Midwestern high schools, where they existed, had begun as union schools in which two or more school districts had consolidated their resources. Such mergers subsequently permitted the union school to open a normal or high school class. The formal establishment of a high school then was the next step. For Wisconsin farmers the implications were clear: Each move on the road to union or high school increased the distance between their home and the school. It weakened what they so proudly called local control of their neighborhood school. In a high school, far removed from their home, the schoolmen would take over. That had happened in the cities and villages; but, if farmers could help it, that would not occur in the country.[15]

Parents who were fearful that their children might succumb to the "barbarism of the frontier" and desired a post-common school education for their children felt perfectly satisfied with the private academies and other similar educational institutions that dotted the Midwest. In them their sons could be prepared for the opportunities opening in trade and business, their daughters could obtain a proper education for their lives as middle-class wives and mothers or, if it should turn out to be necessary, for earning their keep as school teachers. At the century's beginning in the older, settled areas in the East, private academies had already outnumbered public grammar schools. As the decades passed and settlement spread West and trade and business grew increasingly important, entrepreneurs responded to these parental fears and desires and to the wish of many communities for a reputation as a "college town." They introduced a plethora of different institutions: Finishing schools and academies for girls, teacher and other kind of seminaries for both girls and boys, literary and theological institutions, manual labor schools, and institutes and academies of many kinds. Most of these offered a scientific and "useful" curriculum. But for those relatively few parents who wanted a college education for their sons or daughters, they provided preparatory academies and boarding schools. Private colleges added their own preparatory departments. Privately run institutions flourished in every area of post-common school education.[16] Their numerical strength is documented in the reports of the U.S. Commissioner of Education, which show that in 1876, 86.5 percent of the nation's secondary schools were in private hands and attracted as many as 82.3 percent of the country's secondary school population.[17]

The multitude of private post-common school institutions only strengthened the Wisconsin farmers' antipathy toward the public high school and was reflected in its slow growth in the state's rural areas. Before 1868 all of the state's first twelve high schools had been opened in cities. Though the legislature in 1854 and 1858 had authorized the establishment of high schools through district consolidation whenever two-thirds of the district voters agreed, nothing had happened in rural areas. To spur on the reluctant farmers State Superintendent Josiah Pickard persuaded the legislature in 1861 to replace the township organization of district schools with the county superintendency, hoping that centralization would increase efficiency in school organization and administration. But no such result occurred. Not just rural areas, even cities and villages were not always eager to tax themselves for a high school. When after the end of the Civil War Platteville became the host of the state's first public normal school, its citizens congratulated themselves on having been given, at the state's expense, the equivalent of a private academy or public high school.[18] They now saw no need to bother about a high school. In Whitewater, the state's second public normal school began its work in 1868. For nearly twenty years it kept that city from opening a high school of its own. Much to the chagrin of the schoolmen, Wisconsin citizens did not embrace the high school with enthusiasm.

Change came only with the Free High School Law of 1875. That change, however, came mainly in the cities and villages. The law, inspired and promoted by State Superintendent Edward Searing, authorized any town, incorporated village, or city to open one or two free high schools. As a special incentive to rural areas, the law permitted two or more adjoining towns to jointly establish such a school. Two or more adjoining districts could do the same if towns declined to act. When a high school had been maintained for at least thirteen weeks during the year, the state was to furnish school aid at half the instructional costs incurred.[19] Searing had deliberately placed the law's focus on townships and districts instead of on the counties. To appeal even more to the local concerns of the farming population, he had argued that township high schools were to be supplementary to the district schools rather than preparatory for the university. For the financing of the schools he had persuaded the legislators to draw on the state treasury rather than on the more limited resources of the state school fund. Searing had done everything he could to make the law acceptable to rural parents and taxpayers. Still, of the more than one hundred high schools that came into existence in the state in the next five years 70 percent could be found in villages and 30 percent in cities.[20]

In the townships the reaction had been minimal. As the superintendent of Sauk County reported, "the project wherever broached has been looked upon as a new scheme to bleed an already overburdened public."[21] State Superintendent Robert Graham, in office from 1882 to 1887, sought to counteract that sentiment and asked the legislature for authority to supervise the high schools directly, thus circumventing the exclusive jurisdiction of local boards. Though he received that authority in 1883 as well as an additional legislative appropriation two years later, the desired result failed to appear.[22] By 1889 his successor reported that only four towns had opened their free high schools.[23] The township high school remained a problem child. It never could overcome its major impediment: The low population density of rural areas and the resulting inability or unwillingness of local taxpayers to support it. That situation did not change until 1909 when the legislature passed the Union Free High School Law that allowed high schools, including the few township high schools, to merge and draw their students from a wider area.[24]

In rural areas all across the country throughout the nineteenth and into the twentieth century, the professional schoolmen in the state departments of education were locked in battle with their antagonists, "the educators in overalls" and the farmers they represented.[25] When Wisconsin State Superintendent Pickard replaced the township organization of district schools in 1861 with the county superintendency, he had merely pushed the problem up one level. By 1900, when county superintendents were present in all Midwestern states but Ohio, they had taken over the function of the township superintendents. Their assignment was to bridge the gap between the state superintendents and the multitude of local and district school boards. But as in most cases the county superintendents, just as the town superintendents before them, were elected by the farming population from among their neighbors, they found themselves in an unenviable position. They were caught in the crossfire between the professionals on the state level who expected them to be efficient and businesslike and their neighbors who held them accountable to their concerns. As the county superintendents had to hire and fire teachers and draw district boundary lines they had to be politically savvy and able to manouevre among the conflicting local interests. Such agility almost always aroused the suspicion of the professional schoolmen at the state level who distrusted the county superintendents despite, or perhaps because, of their dependence on them. But when the professionals sought to eliminate the county superintendents' positions, as they did in Wisconsin at the turn of the century, they failed. As Wayne Fuller explained, the schoolmen "threatened to deprive the farmers of their right to supervise the education of their children, and [the farmers] would not have it."[26]

The final battle of the defenders of local district control with the schoolmen was the controversy over school consolidation. It took place after the turn of the century primarily in the rural Midwest. For the rural taxpayers and parents, the issues were many and varied. The closing of their district schoolhouses, the loss of control over the selection of their children's teachers, the introduction of "book farming" for boys and "domestic science" for girls, the "wagoning"—as busing then was called—of their children over long distances and poor roads, the discrimination their children often experienced in the new consolidated schools at the hand of nonfarm youngsters, and the new and higher property taxes that soon followed weighed heavily on the thinking of farmers.[27] For a time, they were successful in staving off consolidation. But by the 1950s the decline in the country's farm population had given added impetus to the demands of the schoolmen and consolidation ultimately won the day.[28] As David Tyack put it: American schoolmen were on their way to devise the "one best system" for educating all of the country's children everywhere.

Localism, the defense of local control over schools, was not the only banner under which parents and taxpayers in Wisconsin and other states had fought the growing tide of uniform state regulation. Native born and immigrant Catholics in the rural districts, just like the Catholic immigrants in the New York of the 1830s, resented the Protestant religiosity that pervaded the public schools. They and their fellow Catholics all across the country objected to the use of the King James Bible in the classroom and to having to pay taxes for what they considered Protestant public schools while they received no tax revenues for Catholic parochial schools. They asked that, like in Prussia, public schools be either Protestant or Catholic or, if that was not possible, that their children be allowed to receive Catholic religious instruction from Catholic teachers in public school classrooms. As Catholics failed to gain these concessions from the leadership of the common schools, they began to build a parochial school system of their own. As long as they were willing to pay for it, American Catholics had devised their own system of school choice.

In Wisconsin the children of Catholic and Lutheran immigrants who in 1894 were enrolled in private or parochial schools amounted to one-sixth of all children between the ages of seven and thirteen. There were 279 Catholic parochial schools enrolling 44,669 pupils and 319 Lutheran schools with 20,000 pupils. In addition, in many public schools pastors served as teachers or served on the district boards. German Lutheran immigrants, the majority of whom were organized in the Wisconsin and the Missouri synods, made it a special point to preserve the German language as well as their Lutheran heritage in their parochial schools. Quite often, they succeeded to do so as well in the public schools. And this despite the laws the legislature had passed in 1852 and 1867 that the traditional common school subjects were to be taught in English and that, as a 1869 law specified, teaching in a foreign language was permitted only for one hour per school day.

How this adherence of immigrants to their accustomed ways affected officials of the public school system shows up well in the 1877 report of the Ozaukee County superintendent to the state superintendent:

> While I have no doubt that not as much is done in the English language as ought to be done, and knowing as I do that these Germans keep more school than the Americans . . . the question presented itself in this manner to me: Shall I, by my action, kill these schools, create a feeling against the common school system and cause the establishment of private schools; or shall I take what I can get, knowing that the next generation . . . will work into English entirely . . . Is not

> an educated German better than an ignorant one, even if he is educated only in German?[29]

We have no record of the state superintendent's response, but we know that the use of the German language continued in many of Wisconsin's schools.

The struggle between the defenders of local interests in matters of religion and language and their antagonists throughout Wisconsin came to a climax in 1889/1890. The Bennett Law, passed by the legislature in 1889, made schooling compulsory for all children between the ages of seven and fourteen and compelled instruction in the basic fields to be in English. For parochial schools whose work and certificates were to be recognized as equivalent to that of the public schools, English had to be the instructional language in their religion classes as well. The law caused a firestorm of resentment among immigrant groups. The Republican Party, which had supported the law as a defence of the "sacredness of the little red school house," was soundly defeated at the polls by a coalition of German, Irish, and Polish Catholics as well as German and Scandinavian Lutherans. The law itself was annulled in 1890. In the same year the Wisconsin Supreme Court handed down a decision in the so-called Edgerton case. Catholic parents had sued the Edgerton school board for permitting the King James version of the Bible to be read in public school classes. This, they argued, violated the Wisconsin Constitution, which prohibited sectarian instruction in the public schools. In agreeing with the parents, the Court rebuked the public schools and handed another victory to the partisans of local control.

The Wisconsin victories of 1890, however, did not signify a national trend. On the contrary, by the time the new century had begun, across the country state and national rather than local educational authorities determined the course of public schooling. Again, the clashing philosophies and purposes of parents and professional educators became manifest. Not only in rural Wisconsin but also in cities like Boston and Philadelphia parents had asked for high schools to serve as "people's colleges," that is, less expensive and close-to-home alternatives for residential liberal arts colleges. These people's colleges were to serve the needs of young people and localities for advanced employment training. But the influx of girl students and the tendency in the Midwest to see these schools not so much as alternatives to but as feeders for liberal arts colleges led to a gradual downplaying of their local significance.[30] It did not take long until education professionals, like state legislators and national statesmen before them and ever after, subordinated local interests to state and national concerns.

In large cities and metropolitan areas the rise to power of educational administrators had been prompted by a change in the structure of school government. During the century's closing decades an alliance of businessmen, professionals, journalists, and other leading citizens—the movers and shakers of the Progressive movement—had spearheaded a campaign against inefficiency and corruption in city and public school administration. The results soon became evident. In city after city the ward system of board government was replaced by central school boards whose members no longer represented their ward constituents but, in most cases, were elected at large. While in 1893 in twenty-eight American cities larger than 100,000 inhabitants there had been 603 central school board members—or 21.5 per 1,000 residents—this number had shrunk in 1913 to 264 or 10.2 per 1,000. By 1923 that number had further declined to 7 per 1,000.[31] Consolidation and centralization were the hallmarks of the movement.

The central board members decided on policy for the city's school system in much the same way they did for business and charitable corporations. Yet they left the day-to-day administration of the schools to the city superintendent, an expert with his staff of specialists trained in the "science of education" in university departments of educational administration. As David Tyack put it,

> With centralization and the corporate model in the large cities came the growth of vast and layered bureaucracies of specialized offices, differentiation of patterns of schooling to the specifications of a new "science" of education, Byzantine organization charts, ten of thousands of incumbents protected by tenure, and many people within the city bewildered about how to influence the behemoth that had promised accountability.[32]

With the rise of the city superintendent now complementing the position of his colleague in the state educational hierarchy, the modern educational profession had come into its own.

But as the example of resistance to school consolidation and enforced English language use in Wisconsin's rural schools has shown, the rule of the administrative progressives did not remain unchallenged. This held true for the cities as much as for the countryside. In Rochester, New York, Bishop Bernard J. McQuaid spoke out in the 1870s in support of parental rights to decide about the schooling of their children, and called the city's public schools Protestant and communist. In Rochester, Toledo, and Milwaukee, German parents asked the schools to offer German language instruction, and succeeded in the latter two cities.[33] Parents, organizing themselves during the 1890s in mothers' clubs and unions and parent-teacher organizations, opposed the administrative progressives' drive toward centralization and professionalization. In 1897 they first gained national recognition when they created the National Congress of Mothers. They accepted the often perfunctory invitation of school board members and educators to become involved in the affairs of their public schools. They began to agitate for a wider outreach of the schools into their surrounding communities. They wanted them to become community centers and add vacation programs, school lunches, and playgrounds. In their view, the public schools were to serve their neighborhoods and their special needs. They were not to function as streamlined assembly lines set up to feed the business and industrial demands of city, state, and nation.

Still, the centralization, systematization, and professionalization of public education continued. Its prime moving power lay in the educators' professional organizations, most important among them the National Education Association. This group had begun in the 1870s as an organization of school administrators, teacher educators, college and university presidents. Its most notable achievement during this stage of its career was the issuance of the Committee of Ten Report in 1894, which proposed a common, heavily academic curriculum for the nation's high schools. As two educational historians pointed out, this report was "a crucial first step toward the professionalization of curriculum planning and . . . a direct assault on the control of high school curricula by lay boards of education."[34] In 1918, when the association no longer represented the country's academic elite of college and university professors but the new educational leadership of faculty members of schools of education and of educational administrators and specialists, it continued its determining role of the high school curriculum with its Cardinal Principles Report of Secondary Education. From now on it intended to operate on a national level and to promote there the interests of the education professionals.

This consolidation of institutional power on state and national levels continued into the twentieth century. In the wide spaces of the interior, state administrators set out to counterbalance the fragmentation and isolation of local communities through persuasion and exhortation. In the cities where poverty and slums developed and freed slaves and their children joined the urban masses, educators relied on truant officers and policemen. State laws made education compulsory and overruled local arrangements. The progressive era not only brought much grassroots protest and reform, but it also strengthened the centralizing and homogenizing forces of an efficiency-minded administrative hierarchy. The turn-of-the-century high school disseminated values among the young that reflected the new industrialized and commercialized ways of earning a living. As the schools' managers presided over manifold social, athletic, and academic activities, they came to act as captains of a community's social life.[35] Where once parents and voters had shaped the schools' agenda, now the education professionals directed their course.

Prussia's and later Germany's education professionals were the education *Beamten*. As the nineteenth century wore on, their permanent presence shifted power of the country's schools toward the administrative state and decreased the likelihood that the voices of parents would receive a hearing. Their abiding strength and influence derived from the common education and *esprit de corps* they had received in the humanistic *Gymnasia* and in the universities. That education had been designed, as Hegel had pointed out, to prepare German male youth for careers in state service that "demanded sacrificing the satisfaction of one's independently chosen subjective purposes, but at the same time bestowed the right to find them in, and only in, dutiful service."[36] To a large extent, the immersion of the *Beamten* in a duty-bound, public-interest and nonpartisan *Weltanschauung* enabled the state administration to pursue its measures relatively uninfluenced by fluctuations in the government's policies from liberal to conservative. Even after the conservative reaction had begun in the wake of the Karlsbad decrees of 1819, Prussia's *Beamten* succeeded in preserving at least some of the liberal policies of the Humboldt era.

This was the case with Humboldt's preference for general education over utilitarian training and with his admiration for the liberal pedagogical principles of the Swiss educator, Johann Heinrich Pestalozzi. Humboldt had set into motion a countrywide common general education program for all students and had persuaded the Prussian government to send aspiring teachers to Switzerland in order to absorb Pestalozzi's educational philosophy. Through a program of general education Humboldt wanted to create the possibility for every male student, without being shunted off into vocational training, to attend *Gymnasia* and universities as far as his abilities and means would permit.[37] In the elementary schools Pestalozzianism was to revivify teaching and evoke the students' own creative and imaginative powers. As it turned out, the humanistic training of Prussia's *Beamten* bore fruit. Humboldt's views on general education, while not accepted everywhere in society and government, found enough supporters among state administrators to keep them alive throughout most of the century. The Pestalozzian appeal to students' self-activity and creativity was passed on to future generations of teachers by ardent and capable liberal directors of Prussia's teacher training seminars, even when they, at times, had to ignore and actively counteract the directions of a conservative ministry.

But the country's turn to reaction and conservatism in the wake of the Karlsbad Decrees of 1819 put the proponents of the socially progressive aspects of Humboldt's policies on the defensive. They now came in conflict with administrators who emphasized class- and estate-specific schools and relied on the church as ally of the state. The fear of revolutionary republicanism prompted the government of Friedrich Wilhelm III to restrict the length of school attendance and the content of the curriculum in the country's primary schools, especially those in rural areas.[38] Baron von Altenstein, the minister responsible from 1817 to 1840 for Prussia's educational, religious, and cultural affairs, engaged in a precarious balancing act. He tried to keep alive as much as possible the pedagogical reforms initiated in the Humboldt era while yet he had to conform to the letter, if not the spirit, of the reaction that followed the decrees of 1819.[39]

Both Altenstein and his deputy Johann Wilhelm von Süvern were firm supporters of the famed Prussian alliance of throne and altar. In Prussia, unlike in the United States, churches were not private organizations belonging to civil society. Instead, the Protestant Church was part of the state establishment. Altenstein insisted on the teaching of religion and the practice of daily prayer and worship in the public schools. He implored teachers to remember their position as exemplars and faithfully to accept the performance of religious duties as part of their responsibilities. Süvern wrote that the teaching of religion not only fostered piety, but also was an important means to preserve discipline. Both men thus cemented confessionalism into the public schools and thereby virtually eliminated parochial church schools as viable alternatives.[40] When Süvern recommended to the Theological Faculty of the University of Berlin to send its best students as instructors into the teacher training seminars he argued that these future pastors, having had such exposure to pedagogy, would be in a most excellent position later to serve as superintendents and inspectors of the rural schools.[41] Rural school administrators and Protestant ministers, both civil servants, merged their professional identities. At the royal court in Berlin and in the smallest village, church and state appeared as one.

Yet when in 1819 Süvern drafted a statewide school bill—which was never adopted into law—he sought to moderate the supervisory power of the state church by adding lay members to local school boards. In cities these boards should consist of representatives of the municipal administration, of one priest or pastor, and one or two "house fathers."[42] In the country, the local patron should be added if he contributed financially to the support of the school.[43] Süvern's inclusion of the "house fathers" was prompted by his commitment to a system of national education which, if it was ever to be fully realized, required the attendance of all children in the country's public schools. Because parents of economically well-situated families usually taught their children at home or sent them to private preparatory schools (*Vorschulen*), they and taxpayers without children needed special incentives to support the public schools. Süvern's bill provided these by giving fathers and taxpayers the right to be informed about school policies and instruction, to submit complaints, and to have these accepted and investigated.[44]

Süvern's bill, however, stopped short of allowing parents to freely select their own representatives. He specified that the "house fathers" had to be contributing members of the school society which was responsible for the school's financial support. They were to be duly elected by that society and approved by the provincial school administration. Such approval was to be given only if the candidates were "honest and sensible gentlemen who thought well of schooling and education and were respected by their fellow citizens."[45] This was the first time that fathers, though not mothers, were formally recognized as accepted parts of the country's school system. But as the mechanism for their election made clear, the administration was to have the last word in their appointments.[46]

Limited as Süvern's concessions to more liberal school policies were, they could not sway his conservative colleagues in the ministry. One of these, Ludolf von Beckedorff, protested what he called the artificial egalitarianism of Süvern's proposals. He contrasted it with the natural inequality of human conditions which, he wrote, fostered the bonds of mutual assistance. According to Beckedorff, a liberal education offered in rural schools would only lead to arrogance in members of the lower classes. It would make them demanding and unwilling to carry out the necessary menial work of society. As a result, they would become thoroughly dissatisfied with their lot and dangerous to social peace.[47] Furthermore, Beckedorff complained, Pestalozzian teachers forgot that elementary schools in the country were to teach only the basics of literacy and numeracy. By offering advanced subjects they encouraged their students to leave their communities or their occupations. He therefore censored the frequent attempts of liberal directors in the teacher training seminars to offer subjects like algebra, geometry, surveying, physics, psychology, and mineralogy and to introduce the principles of Pestalozzianism.

Such pedagogy, Beckedorff wrote, "was unwilling to accept authority and, instead, elevated the individual judgment as the touchstone for the correctness of each doctrine. [It] . . . will educate only sophists and doubters and, while it maintains that it fosters independence, will create undisciplined views and arbitrariness of character."[48] The specter of an overeducated peasantry, made rebellious and unmindful of their "proper place in life" by a radical Pestalozzianism, haunted officials like Beckedorff.

By 1822 Beckedorff's views had gained the upper hand. A royal edict of June 15 confirmed that elementary school teachers were sufficiently trained if they were able to teach reading, writing, arithmetic, and religion. Subjects like grammar, geography—at least in so far as it concerned foreign countries—political and natural history, science, and drawing were unnecessary, wrote Friedrich Wilhelm III.[49] Altenstein in turn urged all provincial school officials to stress "not many and diverse subjects, but thorough knowledge of what is necessary and indispensable," and, of course, the basis of all true education: piety, fear of God, and Christian humility.[50] It needs no further comments to point out that parental concerns played little role in the government's directives for the common elementary schools. The opportunity for parents to choose their children's schooling was limited by social class and locality. Their religion determined whether their children would attend a Protestant or a Catholic public school. If the government had its way in either one, their children would be exposed to a bare minimum of literacy and numeracy. As Süvern's bill was never adopted, the "house fathers" were never empowered to submit their complaints to local school administrations. The only relief sometimes came through teachers or local administrators, whether secular or religious, who chose to ignore the governmental policies.

But such opposition to or avoidance of the ministry's edicts was generally ineffective. Moreover, originating among administrators, it did little to voice parental concerns. The same could be said of the organized resistance among teachers that began in the 1820s on a countrywide basis. Teachers had their own battle to fight and objected to their financial dependence on local taxpayers. One of their most articulate spokesmen, Wilhelm Wander, called in 1848 for a national

German common school which, funded by the state, was independent of the vagaries of local financing. Wander wanted well-educated and well-paid teachers who, having had access to university training, could enter on administrative as well as classroom careers. He asked for an elementary school curriculum developed along Pestalozzian principles; professional rather than clerical supervisors; a school that was close to the people and free of confessional bonds; and for a system of continuing education throughout life. Wander's appeal summed up the teachers' desire for professional self-determination. It was met by governmental strictures and disciplinary proceedings. Wander was soon dismissed from office and escaped to the United States.[51] Had his proposals been accepted they certainly would have enhanced the teachers' lot. They would have weakened, perhaps even abolished, clerical school supervision and, by the same token, strengthened the state administrative system. But they would not have improved what Süvern had hoped for: opportunities for public participation in school affairs and parental input and choice. As Wander saw it, local taxpayers and parents did not support teachers in their struggle for professional self-determination. Their participation in school governance would have kept teacher's pay and working conditions on a low level.

The failed revolution of 1848 prompted the Prussian government to assert its full force in bringing the country's educational efforts—and especially the elementary schools—under its full control. The crown blamed the instructors in the teacher seminars for the disaster of the revolution. Addressing the seminar directors, the king stated in 1849: "All this misery that descended upon Prussia last year is your fault alone, the fault of your sophistry, your sacrilegious human knowledge which you offer as true wisdom and with which you have eradicated faith and loyalty in the minds of my subjects and have turned their hearts away from me."[52] A few years later the king's anger translated itself into a set of directives addressed to various school officials.[53] The directors of teacher seminars were reminded that their task was to enable future elementary school teachers to offer "simple and fruitful instruction within the bounds of the elementary schools . . ."[54] The purpose of the elementary schools, so read the third directive, was not to aid "an abstract system or an idea of science," but to serve "the practical life in church, family, occupation, community, and state." The schools were to prepare for this life by being themselves grounded in it and remaining within its boundaries.[55]

The Prussian government's central aim now was to restore and maintain social stability. It declared Pestalozzianism a heresy and ordered elementary teachers to teach loyalty to crown and church. Education was to consist of information and indoctrination that would keep young people safely within the rural world into which they were born. Teachers were told that to encourage change and choice was dangerous and therefore forbidden. They were to cherish the supreme role of religion in the schools' curriculum and to acknowledge the ministers of the church as their spiritual guides and professional supervisors. Parents were regarded as subjects who were to fulfill their duties in the context of the existing social order. Obviously, the government had no sympathy for any desire parents might have for their children's social or economic advance. That would perturb the existing social order. As one German scholar put it, the desired product of the Prussian elementary school was "a human being kept within narrow geographic, economic, and cognitive bounds."[56] Under those conditions parental school choice could not exist.

In Prussia's former Polish territories the government carried out an even more deliberate restrictive school policy toward ethnic minorities. It sought to prevent as much as possible the use of the Polish language in the schools. In Posen teachers of secondary schools were advised by the provincial government not to join a scholarly society whose purpose was to cultivate Polish language and culture. As Polish inhabitants usually were Catholics and German Protestants, the government favored Protestant *Gymnasia* of which in 1860 there were four of them to serve 427,000 Germans whereas three had to suffice for the 975,000 Catholic Poles.[57] It does not need repeating that parental desires for the schools their children should attend was given no hearing.

In Germany, as in the United States, resistance to enforced systematization of public schooling did exist, and there were instances where parents managed to attain for themselves a certain amount of school choice. As Prussia's school policy had shown a decided preference for confessional public elementary schools whose denominational character was set by the religious constituency of the local community, and as the great majority of the Prussian population attended one of the two Protestant denominations or the Catholic Church, the government's policy disadvantaged Jewish parents and parents of other minority religious groups.[58] In many larger cities Jewish parents overcame this handicap. Eager to assimilate their young people into Prussia's bourgeois society while yet having them under the tutelage of Jewish teachers, they managed to have many of their private schools replaced with public Hebrew elementary schools. They were helped in this because Jewish teachers, forbidden to teach in Christian public schools, could practice their profession in Jewish public schools.[59] As a result between 1861 and 1901 the number of Jewish public schools rose from 141 to 244, an increase of 73 percent, and this can be compared with an increase in all other public elementary schools of 48 percent (see table 2.1). But not all Jewish parents who sent their children to public elementary schools sent them to Jewish public schools. In fact, the percentage of Jewish children attending public Jewish elementary schools declined from 37 percent in 1886 to 29 percent in 1901 while the percentage of those enrolled in Protestant, nondenominational, or Catholic Christian schools rose correspondingly from 63 to 71 percent (see table 2.2). The result was that the number of pupils in the existing public Jewish elementary schools steadily declined from an average of 138 in 1891 to 106 in 1901.

Despite the conservative slant of Prussian officialdom there remained liberal administrators who spoke up for the concerns of teachers and taxpayers. Adolph Diesterweg was one of these. Diesterweg had served for twelve years as director of the teacher seminary at Mörs in the Rhineland and for fifteen years as director of the teacher seminary for Berlin's city schools until he was forced into retirement in 1847. He had consistently argued to shift the administrative and supervisory powers over the schools away from the state bureaucracy to the members of local school societies who, as taxpayers, supported the schools financially. He suggested that the school societies be empowered to elect teachers and the members of local school boards and to decide school policies within the framework set by law. Within the classroom teachers should remain free to arrange their work according to their best professional judgment.

Table 2.1 Number of Public Elementary Schools in Prussia, 1861–1901

Year	Protestant	Catholic	Nondenominational Christian	Jewish
1861	16,540	8,082	?	141
1886	23,122	10,061	515	318
1891	23,749	10,154	595	244
1896	24,487	10,725	680	246
1901	24,910	10,799	803	244

Source: Based on Bureau für Statistik der Juden, *Der Anteil der Juden am Unterrichtswesen in Preußen* (Berlin: Verlag des Bureaus für Statistik der Juden, 1905), p. 28.

Table 2.2 Number and Percentage of Jewish Students in Prussia's Christian and Jewish Public Elementary Schools, 1886–1901

Year	Protestant	Catholic	Nondenominational Christian	Jewish
1886	11,451 (32.3)	4,949 (14.1)	5,762 (16.2)	13,249 (37.4)
1887	10,853 (35.7)	4,327 (14.2)	5,704 (18.2)	9,502 (31.3)
1896	9,231 (34.1)	3,857 (14.2)	5,804 (21.5)	8,123 (30.1)
1901	8,640 (36.0)	3,283 (13.6)	5,162 (21.5)	6,937 (28.9)

Source: Based on Bureau für Statistik der Juden, *Der Anteil der Juden am Unterrichtswesen in Preußen* (Berlin: Verlag des Bureaus für Statistik der Juden, 1905), p. 30.

Parents, not members of the clergy, Diesterweg wrote, were to serve on local school boards and to represent the interests of religion. This was a calculated step on Diesterweg's part to separate the competencies of taxpayers from those of parents. Both were to free teachers from supervision by state and church. Taxpayers were to keep the state out of the administration of schools by preserving local control; parents were to free the teachers from clerical oversight in the classroom.[60] Teacher organizations, however, remained ambivalent. In subsequent reform proposals they submitted, they remained silent on Diesterweg's idea to involve the participation of parents and taxpayers in school supervision. Teachers were far more concerned to assert their own desire for autonomy in the classroom and to rid themselves of supervision than to argue for the local taxpayers' claim to a voice in school administration, let alone to speak up for a parental right to have a say in the choice of schools for their children and in the ways teachers manage their classrooms.[61]

Another factor to mitigate the centralizing tendencies of the Prussian and later the German state was the role played by local communities in school administration. As often as not the schools were caught in what one scholar called the struggle between producing society and the administrative state.[62] For much of this period the state's hesitation to finance schools and the in-place presence of local authorities strengthened the hand of parents and communities. A business- and industry-oriented bourgeoisie had encouraged a great variety of city schools, which existed side by side without any specific order or system. Just as colonial Massachusetts localities had protested the General Court's insistence on founding Latin Grammar Schools and had preferred academies instead,[63] so German urban civil society had asserted its interests when it favored its various municipal schools over the state-supported *Gymnasia* and made sure that many *Gymnasia* incorporated classes of the *Realschul* type. The state hesitated to intervene and call for ordered uniformity. Beckedorff, for example, refrained from enforcing compulsory school legislation. While he held that parents were responsible for ensuring the education of their children, he believed that a conservative government had no business to force parents with legislation and police measures to accept their duties.[64] Thus until the failed revolution of 1848 the cities had been left free to arrange their schools as they saw fit, and political parties, municipal and rural commercial and property interests, the churches, teachers, and parents all made their voices heard.

This, however, changed during the 1850s and following decades. A lengthy process of reorganization and reclassification of existing school types and regulations evolved throughout the second half of the century until a comprehensive regulatory scheme emerged. When after the creation of the Second German Reich in 1871 an oversupply of highly schooled young men emerged during the 1880s and 1890s, educational planners devised a vertically differentiated secondary education system of parallel tracks, which led to differentiated occupational qualifications and entitlements. Students were allowed to transfer among the school types only in the lower two grades; their subsequent opportunities for change and their parents' wishes for choice were held to a minimum. As one scholar wrote: "Educational qualifications thus became the primary goals of secondary schooling, and levels of education became the primary criteria of social differentiation in the bureaucracy and in society as a whole."[65]

As parental position in the social hierarchy largely predicted a child's scholastic and subsequently vocational and professional career, the government's prescribed system of school types ensured social reproduction. The accelerating population increase of the century's fourth quarter, the efforts of the new Reich's government to unify and equalize regional and state administrative policies, and the country's industrialization made the process more encompassing and definitive than had been the former hybrid collaboration of state and community agencies. Many parents felt reassured because the system offered clearly marked pathways for their children's career from school to vocation or profession.

Those pathways diverged in a child's tenth year at the time of *Schulwahl.* German parents had to indicate a choice whether they wished their child to remain in a lower school or be transferred to a middle or higher school. If the child was to remain in the lower school a boy would eventually join the ranks of the blue collar working class. If he were to be allowed to transfer he would be expected eventually to proceed to a medium level white collar job or to a professional position. For girls their schooling would largely determine their prospects of marriage and placement in one of the female occupations. Financially and socially able parents of ten-year-old children usually managed to express a preference for their child's advanced school type and, with that, for a subsequent occupational or professional career. The last word, however, remained with the school authorities. Administrators and teachers in both elementary and secondary schools saw it in their professional self-interest to recommend and place children as they deemed best.[66]

The gradually but steadily increasing influence of the national government over the country's schools during the latter half of the nineteenth century and the likewise growing perception of schooling as a national public rather than local private responsibility were prompted primarily by Prussia's and Germany's industrialization and commercialization. In secondary education, this process brought into the open built-in contradictions. On the one hand, the state needed to modernize industry and business and to accommodate the interests of Prussia's propertied bourgeoisie. It therefore often recognized the cities' claims for the various types of modern secondary schools that offered curricula in the natural sciences, modern foreign languages, and commercial and technical subjects. On the other hand, Prussia's governmental bureaucracy, staffed by the highly educated *Beamten* of the educated bourgeoisie, felt obliged to defend the primacy of the humanistic *Gymnasium*, which for most of them had been the portal to positions of eminence in the state. It therefore also often spoke up for the country's educated bourgeoisie whose members feared that the spread of modern secondary schools would lead to overcrowding and thus to a cheapening of secondary and higher education. The country's leading bourgeois class could well have said with Goethe, "Two souls, alas, live in my breast."

Throughout the century's last three decades the struggle between the champions of the modern secondary schools and the defenders of the humanistic *Gymnasium* defined the terms of Germany's school wars. Civil society supplied the driving force behind the push to gain for the graduates of the modern schools entitlements to university entrance and thus access to the professions. Various organizations of modern school teachers and administrators, of professionals working in technical fields, and of instructors in technical universities joined municipal representatives to lobby for the modern city schools. They were opposed by teachers in the *Gymnasia*, university professors, and medical and legal professionals who were reluctant to give up the social prestige attached to their exclusive entitlements as graduates of a humanistic secondary education. While the progressive modernization of the country should have persuaded the government to throw its support to the side of the modern schools, fears of their graduates flooding the labor market and apprehensions about creating an "academic proletariat" prevented this from happening until the end of the century.

The denouement of the German school wars came in the Imperial June School Conference of 1900. The emperor, Wilhelm II, himself had prepared the ground with his address at a school conference ten years earlier in which he took the side of the modernists, the *Realschulmänner*, as they were called. The *Gymnasia* with their humanistic curricula, the emperor had said, had stressed knowledge (*Kennen*) above skills and abilities (*Können*) and had shown little concern whether a student's knowledge fitted him for life. The classical must yield to the national. "We must take German as a basis for the *Gymnasium*," he had said. "We are to educate young Germans, not young Greeks and Romans." A decade later following the conference of 1900 William II abolished the monopoly of the *Gymnasia* and granted entitlements for university entrance to the graduates of two types of modern secondary schools in addition to those of the humanistic *Gymnasia*.[67] The spokesmen for civil society celebrated their victory and pointed to a warning uttered as early as 1878: "In the past," so had said the director of a modern secondary school, "the *Beamte* was the master and the citizen the submissive servant. Today, however, the submissive servant has become a well-off and influential master who prescribes the country's laws which the civil servant will have to execute."[68] The triumphal joy, however, was unjustified. The civil servant remained in his accustomed place of authority. Another way of putting it is to say that an accommodation had been reached between the elites of Germany's educated and propertied bourgeoisie.[69] The state soon showed that, far from having lost authority and having ceded its directive powers over the country's schools to civil society, it had, by granting equal entitlements to modern schools and their graduates, extended its reach over all of the country's secondary schools. Through its educational administrators it made sure that parental *Schulwahl* remained tightly supervised and controlled by local school authorities.

Industrial modernization also spurred the ascendency of state power over the country's elementary and middle schools. Federal contributions to local school expenditures rose from only 5 percent in 1871 to more than 50 percent during the 1890s. This made it possible that tuition and fees, which in 1871 had financed 19 percent of the country's elementary school budgets and had been paid by parents, were abolished in 1888. As a result urban poor schools disappeared and attendance in rural elementary schools rose to nearly 100 percent.[70] In 1872 the Ministry of Culture decreed that from now on elementary schools with more than one classroom were to be considered the norm, and that the number of pupils in rural one-classroom schools must not exceed eighty. Impelled by the demand of industry for literate workers the ministry improved salaries and insurance programs for elementary school teachers. Graduates of the teacher seminaries were granted the privilege of one-year military service instead of the usual three-year term. Middle and vocational schools likewise underwent a government-supported expansion in the years between the foundation of the Second Reich and the outbreak of the World War I. The educational goals of civil society were about to be attained. Yet in the process, the state had solidified its hold over civil society. Now more than ever, the outlook for a *Schulwahl* that gave parents determining powers over their children's academic path was less than encouraging.

Modernization, however, was not the only factor that had prompted the government to extend its influence over the schools. Political considerations also played a significant role. Fear of the social Democratic Party and apprehension over worker militancy lay behind the emperor's order of May 1, 1889. In it he stated that it was the schools' task to convince Germany's youth that the tenets of social democracy contradicted the Ten Commandments and Christian moral teachings and were impossible to put into practice. The schools were to emphasize fear of God and love of country and to pay increased attention to the teaching of modern history and political economy.[71] The same impulse informed the expansion of the *Fortbildungsschulen* (continuation schools), that were to bridge the four-year gap between the boys' graduation from primary school and service in the armed forces. Georg Kerschensteiner, the noted pioneer of vocational education, argued that such schools served the country best if they focused on vocational training which, because it latched onto the self-serving interests of youth, would then automatically lead to a common citizenship education for all.[72]

The government's increasing domination of the country's schools eventually was to include schools for girls, but this was not to happen until the very end of the century. So-called Higher Girls' Schools had come into existence during the eighteenth century as tuition-demanding private establishments. They offered instruction in the basics of elementary schooling, that is, religion and the 3 rs; in "female subjects" such as cooking and sewing intended to introduce the girls to their future duties as housewives; and in cultural accomplishments, such as German language and literature as well as French. Knowledge in the latter fields was to certify the students as properly cultivated future wives of socially respectable gentlemen. By the middle of the nineteenth century parental initiatives had succeeded in many cities to turn their higher girls' schools into community institutions. But the national government had not been in any hurry to include them in its efforts at systematization of curricula and entitlements. Girls, after all, were to be educated as helpmates of men. They were not expected to serve in the civil service or enter into commercial and industrial life. Their schools, therefore, played no role in the affairs of state. They could be left in the care of parents and communities though their directors and upper-level teachers were to be university-trained male servants of the state.

By the 1880s this arrangement came under slowly but steadily increasing attacks. German feminists demanded a *Gymnasium*-level education for women and their subsequent employment as teachers at all levels of the higher girls' schools. In Prussia the first step was taken in 1889 when Helene Lange, a leader of the moderate wing of the German feminist movement, opened in Berlin a program of *Realkurse* for women. These courses offered a modern curriculum of mathematics, natural science, economics, history, modern languages, and Latin. They prepared their graduates for the maturity examination (*Reifeprüfung*) required for enrollment at Swiss universities.[73] An even more decisive step followed in 1893 when Lange converted her *Realkurse* into a program which, by adding instruction in Greek, readied their graduates for the classical *Abitur*, the examination that would permit enrollment in German universities. Two years later the Prussian Education Ministry agreed to allow the graduates of Lange's program to take the *Abitur* at a boy's *Gymnasium*. Subsequently the cities of Breslau and Königsberg inaugurated other such courses as public institutions. When in 1906 the City of Berlin established a public *Abitur* program for girls, Lange's private *Abitur* courses ceased to operate.[74]

The advances in educational opportunities for the higher education of women, however, did not come easily. Officials in the Prussian Ministry of Education as well as many male teachers in the

Gymnasia and professors in the universities opposed and tried to delay them. As one official wrote in 1898, gymnasial education was meant for girls mature and capable enough to come to well-thought-out-decisions, not for twelve-year-olds. Besides, he added, if *Gymnasia* for girls became commonplace, they would cheapen the accomplishments of the higher girls' schools "which did not turn their students into competitors of men but into helpmates, not into scholars or learned blue stockings, but into capable German housewives."[75] It is not surprising, then, that German education officials favored the *Frauenschulen* which did not send their graduates into the *Lycees* and *Studienanstalten*.[76]

Women, however, continued to prod the government to extend to girls, their schools, and their teachers, the same rights and entitlements that had been granted to boys. In 1908 a ministerial directive announced the government's first concessions, without, however, granting full equality. Higher girls' schools whose curriculum would now meet government-set gymnasial standards were to be known as *Lycees* and were to be acknowledged as institutions of higher learning. They were permitted, but not given the right, to be presided over by a female director. Women were given access to the *Abitur* and, with certain restrictions, access to university study.[77] The results of these changes are reflected in the statistics for Prussia's higher schools for women between 1901 and 1911. The 213 public and the 656 private higher girls' schools in 1901 had enrolled 126,490 students. The 231 public and 212 private *Lycees* in 1911 enrolled 144,006 students to which one would have to add the students of more than 400 private schools that did not qualify as *Lycees*.[78] The numerical increase of women preparing themselves for life or university study in institutions of gymnasial education was considerable. Full equality with men in the matter of entitlements to careers in all professional fields, however, eluded them during the *Kaiserreich*. It came to be theirs, together with their right to vote, finally during the early years of the Weimar Republic. By then, the administration and regulation of women's education had moved from the agencies of civil society to the organs of the state.

Chapter 3

School Governance and School Choice 1900–1950

During the first half of the twentieth century the professionalization of American public education moved steadily ahead and made narrower the areas left for parental school choice and participation. Overall policy for public education was set in state legislatures and increasingly by the federal government. It was administered and supervised by educational professionals in state departments of public instruction and the federal Department of Education. These professionals were supplemented by state, county, and city superintendents, most of them trained in university departments of educational administration. In the decades after 1900, they brought a measure of uniform, countrywide administrative educational practice to the school districts of the United States.

When these overwhelmingly male professionals confronted the widely varying sentiments and criticisms of local school boards and parents, they carried out their work under sometimes trying circumstances. They were well aware that under the American system of reliance on local control in public education they depended on the goodwill and support of local taxpayers and parents for success in their work. As Raymond Callahan in his *Education and the Cult of Efficiency* described it, they sought to gain that goodwill and support by shoring up their professional status and their influence in local politics. To that end they relied on the academic road to influence and prestige. They enrolled in graduate programs of educational administration which, in their eventual development, led to the doctorate in education, the Ed. D. degree.[1] They also linked that academic training to the then dominant trend in American life, the pursuit and worship of business. They attempted to gain expertise in business methods, cost accounting, public finance, and scientific management.

In that endeavor they were guided by their instructors in the departments of educational administration of the nation's large universities, men like Frank Spaulding at Yale, John Franklin Bobbitt at Chicago, Edward C. Elliott at Wisconsin, George D. Strayer at Teachers College, and Elwood P. Cubberley at Stanford. These "captains of education," as Callahan called them, created school administration as a profession distinct from teaching. Its separate academic curriculum took its inspiration from business rather than from pedagogy or philosophy. One of their students, William Eastbrook Chancellor, put the matter succinctly: "One cannot be both a thoroughly experienced teacher of youth and a skillful manager of the business of city school systems. . . . The truth is that a new profession has arisen within the old one of teaching."[2] Whether, as Callahan wondered, "administrators were recognized by teachers or laymen as experts or had achieved a degree of professional status in other than their own eyes is difficult to determine. But [Callahan went on], there is no question that by 1918 administrators had followed the authoritarian role of the manager in industry and had applied it in their school systems."[3] There also can be no question that educational administration as a graduate department of research and instruction—a discipline unknown in German universities—had achieved a permanent place in American universities.

No matter how well trained academically and how well attuned to the ways of business these administrative professionals were, their dependence on the often unpredictable moods of local taxpayers prevented them from equaling the power wielded by the *Beamten* in the German school system. But their influence over developments in America's public school system was nonetheless considerable. As I pointed out in chapter 2, the public school administrators and their faculty colleagues in university schools of education managed to take over the leadership of the National Education Association from the elite of college professors and presidents that had set the tone during the latter half of the nineteenth century. Through their professional organizations and in their home school districts these new leaders of American public education set out to shape the administration of the public schools after the model of American business. In his district the local superintendent as professional, served as the chief executive officer of public education. When in their desire to appear as good businessmen the members of school boards urged the superintendent to make efficiency the guiding measure of educational decisions, they came to resemble shareholders; teachers, lacking the university training in educational administration, were reduced to the status of employees; and students and parents supplied the customers.[4] A new type of professional educator stood ready to challenge the older tradition of lay control over the public schools. In the words of David Tyack and Elisabeth Hanson, the "leadership in American public education had gravitated from the part-time educational evangelists who had created the common-school system to a new breed of professional managers who made education a lifelong career and who were reshaping the schools according to canons of business efficiency and scientific expertise."[5]

The drive toward centralized administration of the public schools penetrated rural as well as urban areas and everywhere began to replace local homegrown with professional control. By the end of the nineteenth century parental direction, as it had been effectively brought to bear on rural schools through local boards now had to work its way through the township system and the county superintendency. In 1897 the National Education Association's Committee of Twelve promoted rural school consolidation. In its wake, centralization began to take command. In rural Delaware, to cite one example, the State Department of Public Instruction had by 1920 come to regulate and supervise public education statewide. That administrative oversight extended directly to superintendents, principals, and teachers. When faced with parental involvement in school affairs and with parental complaints about local school facilities or discipline cases, the state superintendent channeled these through organizations like the Parent Teacher Associations. Their diverse membership, the superintendent correctly surmised, would in most cases prevent unanimity and neutralize potential disruption of his work. Whether in urban or rural areas, in most cases parental participation in school affairs stood little chance to deflect administrative direction.[6] On the contrary, as we shall see later on, the new professionals enlisted parents in their public relations efforts to boost the image of the public schools as an enterprise run on business principles.

It would be a mistake, however, to assume that the rise of the professional managers proceeded without hindrance and protest. As I pointed out in chapter 2, women in mothers' clubs, unions, and parent-teacher organizations had begun in the 1890s to demand a voice in school politics. They asked for closer home–school relations and urged that the schools become more actively involved in the life of their neighborhoods. As many of them before their marriage had been teachers or were now wives of teachers, they sought to ally themselves with women teachers in seeking to strengthen bonds between school and home. Across the country they became prime movers in the progressive education reforms of the period. They agitated for and helped to introduce into public education vacation schools, penny lunches, playgrounds, and programs in manual training, domestic science, and vocational training.[7]

Women were not alone in their efforts to counteract the centralizing and business efficiency-minded rule of the professionals. In many American cities they shared their immediate aims with male socialist, populist, and trade union members. In Milwaukee, socialists gained their first representative on the school board in 1909 and sponsored the use of schools as community centers, the distribution of free textbooks, and the presentation of public lectures. They agreed with the women reformers in their immediate aim of making the schools more responsive to the needs of parents and the surrounding community. But, remaining faithful to their socialist convictions, they differed from the liberal women reformers in their ultimate hope of overcoming the capitalist order. As William Reese put it: "Grassroots progressivism . . . had its middle-class and feminine as well as working-class and Socialist roots, growing together in the 1890s like entangling vines that crossed but did not always join."[8]

By 1919 the various reform initiatives that had developed during the preceding decades and had become known as "the new education," found their institutional expression in the Progressive Education Association. Founded on April 4, 1919, in the nation's capital, this association was meant to gather parents and teachers who had become dissatisfied with what they considered the numbing and unimaginative routine of most public schools and who wanted to make the schools vital centers of community life. Stanwood Cobb, a teacher who took the initiative in founding the organization, stressed its intended nonprofessional character. It was, he said, "for the most part, a handful of nobodies, educationally speaking. . . . Our support came from the lay public and a few outstanding liberals in education."[9]

Alas, ten years later a distinct professional tone had come to color the proceedings and pronouncements of the association. Progressive educators in public schools who had embraced practices of the "new education" felt pressed to justify and defend them against critics and colleagues among mainstream professionals who, ironically enough, also considered themselves to be progressives, albeit of the managerial-administrative type. Invariably the "new education" progressives fell back on the profession's standard proceedings of commissions and committees staffed by themselves. Inevitably professionals became the association's spokespersons.[10] This tendency to rely on the trappings of professionalism as well as the unavoidable occasions when educators found themselves to be both partners and mentors in their relationships with parents made it difficult for lay members of the association to feel regarded and treated as equals.[11]

Thus the association began to lose the popular acclaim it had gained among the lay public. When George Counts in a 1932 address challenged the schools "to build a new social order" and asked progressive education "to become less frightened than it is today of the bogies of imposition and indoctrination," his militant tone did not help his cause.[12] Neither was the association aided by the incessant academic debates that followed Counts' call to arms nor by the stigma of radicalism that his appeal had foisted on the group. The association's leaders did nothing to shore up their sagging popular support. When in 1950/1951 a storm broke out in Pasadena over the dismissal of Superintendent Willard Goslin and he and progressive educators across the country were being accused of watering down traditional school subjects with "fads and frills" and of replacing patriotism with internationalism and socialism, they were unable to rally the public to their side.[13] "Within their own councils," wrote Lawrence Cremin, "they devoted themselves increasingly to internecine warfare. . . . More and more they responded to criticism in the ringing rhetoric of self-justification, and in the pages of their own learned journals. The faithful were roused and reconverted, but meanwhile large uncommitted segments of the public listened attentively to the critics."[14] The progressive professionals had lost touch with their base of lay support.

It also did not help the cause of the "new education" when in the large urban centers in the East many of the innovations that reformers had introduced were taken over by public school authorities and became part of the regular schools. Public schools now added visiting teachers, vocational counselors, nurses, and social workers to their staff. The educators' concerns ranged from parent education and the "Americanization" of immigrant families to social welfare issues of low-income families. But this school–home relationship was no longer a matter of mutual exchanges. It had become decidedly one-way. Teachers assumed the role of instructors of those whom the professionals considered to be "the real problem," parents in immigrant, minority, and working-class homes.[15] In 1910 in New York City a public school administrator declared: "The public school teaches how to live. It enforces its lessons of hygiene, of cleanliness, order, and prevention of diseases in a way which is wholly new to the immigrant, and many a home has been cleaned up owing to the combined influence of seven or eight children, which is a far greater influence than the authority of the parents may withstand."[16] When confronted with professional authority, parents had to yield. This was the new dispensation.

All the while the usurpation of progressive innovations continued. Across the country, public school systems and municipalities took over vacation schools, and these lost much of their experimental nature. Domestic science and manual education classes became part of the regular curriculum, and participation was no longer voluntary. In 1918 the National Education Association announced in their curricular blueprint for the American high school, the *Cardinal Principles of Secondary Education*, its interventionist interest in the home environment of their students when they made "worthy home membership" one of the goals of classroom instruction. If "the new education" had intended to use the public schools for the purpose of strengthening a common community life, the public schools now saw their mission as one of instructing and converting minority, ethnic, and working-class parents to the way of the white middle class. As William Reese observed, the public schools lent to many of the once progressive innovations "their heavy emphasis on textbooks, traditional discipline, rote memorization, and teacher authority. In this respect," he added, "the more the schools changed, the more they stayed the same."[17]

At the heart of this transformation of the progressive impulse lay a reconsideration of the school–parent relationship. In the early days of American common schooling parents had wielded considerable influence over the governance of school districts. The parent was the most directly concerned taxpayer and habitually exercised stingy but effective oversight over schools and teachers. The usually female teacher would encounter a parent as father or mother of her pupils or as a male member of the district school board where he served as her employer and supervisor. In either case the parent was to the teacher a person of authority whose disapprobation or anger she was ill advised to provoke. With the rise of the modern educational profession, however, both parents and teachers lost influence to the professional educational manager. The administrator became the teacher's employer and supervisor whose goodwill and approbation the teacher had to court. The teacher, conscious of her own professional training and dependent primarily on her supervisor's opinion of her, now had less reason to stand in awe of a parent. She was less willing to listen to parents and their complaints or wishes. "Parents," wrote William Cutler, "may not be the teacher's natural enemy, but they are usually unwelcome in the classroom."[18] Their formerly dominant role faded.

The professional disregard for lay opinion as unscientific or uninformed came to pervade all parts of the educational establishment. By mid-century it could be found in the public debate over the American high school and over the "educational services" it was to provide for youth. James Bryant Conant, former president of Harvard University and ex-ambassador to the Federal Republic of Germany, presented a good example of this attitude. By mid-century he admonished critics and parents who had spoken out in favor of a strong academic program in their children's high school. These parents, Conant wrote, not only presented "a problem to counselors, principals, and teachers" but also had made "a suggestion dangerous to the security of our nation."[19] Conant did not shy away from implying that parental concerns, understandable as they were, came close to being unpatriotic. When confronted with the recommendations of the nation's educational experts, parents should have to hold their peace.

The belittling attitude professionals displayed toward laymen and laywomen who opposed them or disagreed with them did not prevent them from appealing to the lay public to volunteer for service to the public schools. They needed supportive members on school boards and in parent-teacher organizations. In that effort class and gender played a role. Men usually were enlisted in the policy-setting and administrative positions on school boards, and women took on the social, public relations, and parent education functions assigned to the PTAs. But these American parent-teacher associations and similar organizations of the 1920s did not endorse political, religious, or economic causes. They did not see themselves as pedagogical or social reformers. They were not to become instruments for change in the hands of educational nonprofessionals. The professionals, for their part, as William Cutler put it, "came to regard parents as just another element in the schools' constituency. Self-confidence gradually turned to arrogance among these instructional leaders, transforming the idea of an equal partnership between the home and the school into the bureaucratic concept of professionally managed relationship."[20] In that relationship parents were to function as helpmates of the education professionals. They were to be supporting pillars of the public school establishment.

The National Education Association was the leading organization through which schoolmen, and eventually also school women, brought their influence to bear on the nation's educational policies. But it was not the only one. In 1935 the National Education Association together with the American Association of School Administrators launched the Educational Policies Commission. Its twin purposes were to lobby for funds for public education and to insure that such funding, especially when it came from federal sources, did not endanger or interfere with professional control over the public schools. What had spurred the schoolmen to this venture had been the creation of the Civilian Conservation Corps (CCC) in 1933 and of the National Youth Administration (NYA)in 1935, two New Deal agencies that were to attack the problems of youth unemployment and vagrancy. As the Civilian Conservation Corps and the National Youth Administration operated outside of the Federal Education Department—the Civilian Conservation Corps was administered by the War Department and the National Youth Administration by the Works Project Administration—educators viewed the two agencies as directly competing for federal funds with the public schools. They now bent all their efforts to push back this unwanted intrusion of federal influence and to solidify their own hold over the schools. In these struggles, little was heard from parents and students. The educators' professional self-interest stood at the center.[21]

The Educational Policies Commission soon issued a number of statements which, in the words of David Angus and Jeffrey Mirel, implicitly assumed "that, in a democracy, the aims and purposes of the schools are what the professionals say they are, not what the electorate decide they should be."[22] And what educators then projected as the aims and purposes of public schooling was an education that was to offer something for everybody. In the commission's *Education for all American Youth* of 1944 they charged public education with preparing all youth for vocation, citizenship, and family living; with making them appreciate health, art, and culture; and with teaching them how to become intellectually curious, to pursue happiness, to think rationally, and to appreciate ethical values. Never loath to extend their reach and influence, the authors proposed that public schooling should be extended to include grades thirteen and fourteen.[23] In that they were later seconded by James Bryant Conant who in his *Slums and Suburbs* of 1961 suggested that public schools assume responsibility for educational and vocational guidance of out-of-school youth up to the age of twenty-one.[24] Six years later, after Congress had passed the first enabling legislation for federal aid to education, Conant reaffirmed his belief in this policy and blamed the paucity of well-functioning comprehensive high schools on the unwillingness of schoolmen to ask for state and federal assistance. The absence of nationwide common practices, "let alone a national policy," made it impossible to assure a common educational platform, he wrote.[25] The professional schoolmen's ambition to create an all-encompassing nationwide public education system had stopped short of reaching its full extent.

The common educational platform with common practices, such as Conant and many public schoolmen desired, did not exist in 1967 and had not existed in earlier decades and centuries. From its beginning in the colonial period to the twentieth-century American education has been characterized by initiatives of both governments and civil society. It has offered parents the opportunity to choose between public and private schools. State governments supported the revival of the public common schools and sponsored state universities. The federal government stimulated higher education with the Land Grant Act of 1862, and congressional efforts during the post–Civil War decades sought to make common school education compulsory for every child and youngster. Various sectors of civil society, however, had always been responsible for educational opportunities in primary, secondary, and collegiate education.

In American secondary education 32 percent of the students in 1890 attended private institutions, though that percentage decreased steadily until in 1900 it stood at 17.6 percent. During the next fifty years these enrollment percentages further declined, going through a marked trough during the depression of the 1930s and reaching a low of 6.5 percent in 1940 before they rose again to 10.5 percent in 1950. As table 3.1 shows, enrollments in private elementary education increased from 7.65 percent of total elementary school enrollments in 1900 to 12.26 percent in 1950. The combined percentage enrollments of private secondary and elementary education moved up from 8.02 percent in 1900 to 11.86 percent in 1950. While American parents came to rely less on private secondary schools, they increasingly exercised their right to choose private elementary schools for their children.[26] As in previous centuries, American education remained on a double track.

Due to the absence of complete and reliable statistics it is impossible to state precisely the distribution of enrollments between

Table 3.1 Enrollment Totals and Percentages of U.S. Public and Private High and Elementary Schools, 1900–1950

Year	A	B	C	D	E	F	G	H	I
1900	16,855	1,352	8.02	630	111	17.62	16,225	1,241	7.65
1910	19,372	1,558	8.04	1,032	117	11.34	18,340	1,441	7.86
1920	23,278	1,700	7.30	2,414	214	8.86	20,864	1,486	7.12
1930	28,329	2,651	9.36	4,740	341	7.19	23,589	2,310	9.79
1940	28,045	2,611	9.31	7,059	458	6.49	20,985	2,153	10.26
1950	28,492	3,380	11.86	6,397	672	10.50	22,095	2,708	12.26

Notes

A: Total of Public and Private High and Elementary School Enrollments in 1000s
B: Total of Private High and Elementary School Enrollments in 1000s
C: Percentage of Private High and Elementary School Enrollments
D: Total of Public and Private High School Enrollments in 1000s
E: Total of Private High School Enrollments in 1000s
F: Percentage of Private High School Enrollments
G: Total of Public and Private Elementary School Enrollments in 1000s
H: Total of Private Elementary School Enrollments in 1000s
I: Percentage of Private Elementary School Enrollments

Source: Compiled from National Center for Education Statistics, Digest of Education Statistics 1997, Table 3.

religious and nonreligious private schools during the first half of the century. The annual reports of the U.S. Commissioner of Education, however, do give us some information for the years from 1900 to 1915 about the country's private high schools. As presented in table 3.2, the most remarkable development in this group was the decline in the number of secular schools and the increase in the number of schools of religious sponsorship. Within the short time of five years the number of secular schools decreased by roughly one third from 1,033 to 662, while that of schools of religious sponsorship rose by almost 70 percent from 945 to 1,586. It also appears that while in 1900 the average student to school ratio was nearly the same with 55.3 students per school in the secular schools and 56.7 in the religious schools, by 1915, when only three out of ten private high schools were administered under secular sponsorship, these secular schools now each had to accommodate an average of 77.4 students.

The commissioner's reports also allow us to infer from the statistics some of the reasons why parents may have preferred to choose

Table 3.2 Number and Percentage of U.S. Private Secular and Denominational High Schools and Their Students, 1900–1915

Year	A	B	C	D	E	F	G	H	I	J
1900	1,978	1,033	52.22	945	47.78	110,797	57,173	51.60	53,624	48.40
1911	1,979	699	35.32	1,280	64.68	130,589	50,255	38.48	80,334	61.52
1914	2,199	710	32.29	1,489	67.71	154,857	53,528	34.57	101,329	65.43
1915	2,248	662	29.45	1,586	70.55	155,044	51,215	33.03	103,829	66.97

Notes

A: Number of Private High Schools
B: Number of Private Secular High Schools
C: Percentage of Private Secular High Schools
D: Number of Denominational High Schools
E: Percentage of Denominational High Schools
F: Number of Students in Private High Schools
G: Number of Students in Private Secular High Schools
H: Percentage of Students in Private Secular High Schools
I: Number of Students in Denominational High Schools
J: Percentage of Students in Denominational High Schools

Source: Based on the reports of the U.S. Commissioner of Education for 1899–1900, 1911, 1914, and 1916.

Table 3.3 Comparative Data Concerning U.S. Public and Private High Schools 1900–1914

Year	Percentage of students in College Preparatory Courses		Student to teacher ratio		Boys only		Girls only		Coeducational	
Year	**Public**	**Private**	**Public**	**Private**	**Public**	**Private**	**Public**	**Private**	**Public**	**Private**
1900	11	32	25.5	10.9	11.1	88.9	5.5	94.5	84.1	15.9
1911	6	17	21.8	10.8	8.3	91.7	4.7	95.3	91.1	8.9
1914	5	15	21.0	11.1	7.3	92.7	3.4	96.6	92.2	7.8

Source: Compiled from the 1899–1900, 1911, and 1914 reports of the U.S. Commissioner of Education.

private rather than public high schools for their children to attend. Given the fact that, as I show in table 3.3, about three times as many students in private high schools attended college preparatory classes than did students in public high schools, the emphasis on an academic curriculum in the private schools seems evident. Parents no doubt also preferred private schools because of the relatively low

Table 3.4 Number of U.S. Church-Sponsored High Schools and Their Students, 1900–1917. Major Denominations only

Denomination	1900	1911	1914	1916	1917
Roman Catholic	361/15,872	719/35,757	846/49,095	975/56,182	981/58,327
Methodist, including Methodist Episcopal South	103/8,385	104/9,043	116/10,586	110/9,550	109/9,163
Baptist	96/7,173	100/8,743	112/9,665	105/7,439	101/7,359
Episcopal	98/5,145	79/5,229	109/6,774	99/6,389	85/5,499
Presbyterian	93/4,574	63/3,915	63/4,027	65/3,734	68/3,519
Lutheran	32/2,032	51/3,483	56/3,729	57/3,881	56/3,774
Friends	55/3,428	44/2,841	37/2,686	36/2,444	34/2,392
Congregational	51/2,671	31/1,809	35/2,494	31/2,231	29/2,084

Source: Compiled from the 1899–1900, 1911, 1914, 1916, and 1917 reports of the U.S. Commissioner of Education.

student to teacher ratio which, for the years under consideration, remained twice as low as that in the public schools. Finally, when nine out of ten public high schools were coeducational, parents, who preferred for their children to attend single-sex schools, found these to be much more readily available among the private schools.

The increase in the numbers and in the percentage of religiously sponsored private high schools during the century's second decade also shows that religion itself played a major role in drawing students to these schools. Among the various denominations, the Roman Catholic Church had taken the lead. It built on the head start it had gained in the nineteenth century.[27] As listed by the U.S. Commissioner of Education (see table 3.4), in 1900 it enrolled 15,872 students in its 361 high schools. With 29.6 percent of the 53,624 students attending religious secondary schools it constituted the single largest denominational group. Except for the 75 students of two Jewish parochial schools who were listed for the first time in the Commissioner's Report of 1914, and the 248 students who were listed in 1917 for three Jewish schools, the remaining 37,752 students in 1900 were Protestants of various denominations. Methodists, including those belonging to the Methodist Episcopal Church South, led with 103 schools and 8,385 students, followed by the Baptists with 96 schools and 7,173 students, the Episcopalians with 98 schools and 5,145 students, the Presbyterians with 93 schools and

Table 3.5 Establishment of U.S. Private Schools by Time Period. Before 1904, 1904–1953, 1954–1993 in percent of those in existence 1993/1994

Affiliation	Before 1904	Between 1904 and 1953	Between 1954 and 1993
Catholic	21	45	34
Episcopal	13	25	62
Friends	39	16	44
Seventh Day Adventist	6	31	62
Jewish	0	15	84
Christian	9	25	65
Lutheran	21	19	60
Average for religious schools	16	25	59
Montessori	0	3	96
schools for exceptional children	6	13	81
National Association of independent Schools	22	35	43
Other private schools	2	7	90
Average for secular schools	8	15	78

Source: Based on National Center for Education Statistics, *Private Schools in the United States: A Statistical Profile, 1993–94*, Table 1.3, p. 59. Table 11.5.

4,574 students, the Society of Friends with 55 schools and 3,428 students, the Congregational churches with 51 schools and 2,671 students, and the Lutherans with 32 schools and 2,032 students. By 1917 the Lutheran schools had overtaken the Presbyterians, Friends, and Congregationalists in their enrollment figures. As before, the Catholics with their 58,327 students contributed the largest share of students enrolled in denominational high schools. It now amounted to 56 percent.

In the three decades that followed American parents could choose from a great variety of private schools for their children, provided, of course, that such schools were available in their neighborhood and that parents could afford the expenses for tuition and, if that was necessary, for boarding the student. As table 3.5 shows, the opening of new religiously sponsored schools outpaced that of private schools of secular sponsorship. Among the religiously sponsored schools

Catholic institutions continued to be in the lead, followed by those of Seventh Day Adventists, Christian-Reformed, and Episcopalians. If parents were interested in secular schools, those of the National Association of Independent Schools offered the largest selection of coeducational and sex-specific day and boarding schools. Then there were the country day schools of which in 1937 there were one hundred in existence in the United States and whose essential elements, according to Otto F. Kraushaar, were "a full day program of academic and extracurricular activities, and . . . close home ties with full involvement of parents."[28] Parents could also send their children to schools that followed specific educational philosophies, such as the Rudolf Steiner or Waldorf schools, Felix Adler's schools of the Society for Ethical Culture, and the various teacher- or parent-originated schools commonly grouped together under the progressive label and so well described by Lawrence Cremin in his *Transformation of the School.*[29]

Parents and guardians who sought legal confirmation of their right to send their children to private schools, whether religious or secular, could breathe a sigh of relief in 1925 when the U.S. Supreme Court affirmed that right in the case of *Pierce v. Society of Sisters.*[30] The decision, often referred to as the *Magna Charta* of the private school, came in response to appeals from two corporations, one a religious foundation supporting schools and orphanages, the other the board of a military academy. The target of the appeals was an Oregon law that required every parent to send their children between eight and sixteen years to a public school. The Court held that the law would bring about the destruction of private primary schools and that this was impermissible under the due process clause of the Fourteenth Amendment. The justices further ruled that the Oregon law "unreasonably interferes with the liberty of parents and guardians to direct the upbringing and education of children under their control." They did not mean, they said, to strike at the power of states

> reasonably to regulate all schools, to inspect, supervise and examine them, their teachers and pupils; to require that all children of proper age attend some school, that teachers shall be of good moral character and patriotic disposition, that certain studies plainly essential to good

> citizenship must be taught, and that nothing be taught which is manifestly inimical to the public welfare. . . .

But they meant to rule against a state's claim to exercise an educational monopoly through its public schools.

> The fundamental theory of liberty [the Court continued] upon which all governments in this Union repose excludes any general power of the state to standardize its children by forcing them to accept instruction from public teachers only. The child is not the mere creature of the state; those who nurture him and direct his destiny have the right, coupled with the high duty, to recognize and prepare him for additional obligations.

Parents and students as well as private school proprietors, sponsors, and teachers could take heart. While state legislatures and state officials were confirmed in their power to regulate all schools and to enforce compulsory education laws, they were barred from requiring that all children attended public schools only.

For the next twenty-five years, spanning the depression decade of the 1930s and the years of World War II, the Court continued to find the interests of private school parents protected under the provisions of both the Fifth and the Fourteenth Amendment. In a 1927 case the owners, teachers, parents, and children of foreign language schools in Hawaii complained that territorial legislation "would deprive them of their liberty and property without due process of law." The Supreme Court held that the Fifth Amendment applied to the federal government and its agencies just as the Fourteenth Amendment applied to the states and sustained a district court injunction that forbade enforcement of the legislation.[31] In 1930 in *Cochran v. Louisiana State Board of Education* the Supreme Court rejected the contention that taxation for the purpose of purchasing and distributing school books free of cost to the state's school children, including to those in private schools, constituted a taking of private property for private purposes. The legislation, the Court argued, did "not segregate private schools, or their pupils, as its beneficiaries." Furthermore, in what came to be known as the child benefit theory, the Court pointed out that it was not the schools but the children and the state who benefitted from the legislation.[32] Thirteen years later the Court barred local authorities from compelling the children of Jehovah's Witnesses to salute the American flag in their schools. It found that the First Amendment was designed to protect the "sphere of intellect and spirit" that local authorities here had invaded. "No official, high or petty," said the Court, "can prescribe what shall be orthodox in politics, nationalism, religion, or other matters of opinion or force citizens to confess by word or act their faith therein." In this opinion the Court turned to history and pointed to the ultimate futility of other attempts to compel unity. The examples the justices cited were the Roman attempt to stamp out Christianity, the Inquisition, the Siberian exile, and the extermination campaigns conducted by totalitarian regimes. Referring to those atrocities then occurring abroad and the legislation compelling the flag salute at home, the justices added: "[T]he First Amendment to our Constitution was designed to avoid these ends by avoiding these beginnings."[33]

In two cases, decided in 1948 and 1952, the Court dealt with yet another issue of importance to parents who were greatly concerned with the kind of religious instruction their children were or were not receiving in public schools. It concerned the manner in which public schools permitted release time for religious instruction.[34] In the first case, *McCollum v. Board of Education*, the Champaign, Illinois, Council on Religious Education, a voluntary association of Jews, Roman Catholics, and Protestants, had been permitted by the Board of Education to offer religious instruction through ministers of the various faiths in public school classrooms to children who had received parental permission. The Court found that this arrangement was "a utilization of the tax-established and tax-supported public school system to aid religious groups to spread their faith. And it falls squarely under the ban of the First Amendment (made applicable to the States by the Fourteenth) . . ." In the second case, *Zorach v. Clauson*, 343 U.S. 306, City of New York parents could request that their children be released from public school to attend religious instruction at a religious center. That program did not take place within the public schools and it did not require the expenditure of any public money. "Here," the opinion held, "the public schools do no more than accommodate their schedules to a program of outside religious instruction." To say that such accommodation violated the separation of state and church would be to "find in the Constitution a requirement that the government show a callous indifference to religious groups" and to read into the Bill of Rights "a philosophy of hostility to religion." No such requirement and no such philosophy existed. Accordingly, the Court found the release time as practiced in New York City acceptable under the meaning of the First Amendment.

During the first half of the twentieth century the tension that had been building between the seemingly inexorably growing domination of the American public school system by the professionals and the unease and protests of diverse sets of parents and teachers came to a head. The professionals had found their voice in the National Education Association and the Educational Policies Commission. Protesters, whether parents, teachers, Socialists, or trade union members, rallied in the Progressive Education Association and founded the various schools of the New Education Movement. The *Pierce* decision of the U.S. Supreme Court then laid the legal basis for the right of parents to send their children to private schools. Those parents who chose that option, a minority whose children never exceeded more than 12 percent of the total public and private high and elementary school enrollment between 1900 and 1950, did so for many and varied reasons. Many preferred for their children an education in schools that emphasized religious instruction or promised a greater emphasis on academic, artistic, or experimental pedagogy. Others sought schools that permitted or encouraged parental participation in setting school policies or provided a lower student to teacher ratio or offered a single-sex setting. While the various strands of protesting parents and New Education teachers did never match the political strength of the public education professionals they had nevertheless succeeded in stalling the professionals' drive for a public school monopoly.

In Germany public schooling at the beginning of the twentieth century had come increasingly under the control of state authorities. It has remained there ever since. To the great disappointment of Germany's liberal, democratic, and revolutionary teachers and educators, even the military defeat of the German Empire, the turbulent months of the November revolution of 1918, and the debates over the constitution of the Weimar Republic did not bring fundamental change to German schooling. Though the government bureaucracy experienced great strains and stresses and in city-states like Hamburg teachers gained power temporarily at the expense of local administrators, by the time the debate over the constitution had ended, its school clauses reflected a compromise among the various political parties which had, among themselves, represented the interests of state and local school administrations, organized teachers, and churches. In that debate the churches had raised the claims of parents to be heard. They had represented them, not to register parental demands for participation in school administration and for a voice in determining the academic future of their children, but to uphold the maintenance or establishment of confessional public schools.

The Weimar Constitution's general direction for school policy declared that public institutions were to provide the education of youth through the cooperation of the Reich, the *Länder* (states), and localities.[35] The states were to exercise supervision over these institutions and could, if they wished, allow localities to participate. The supervising work itself was to be carried out by full-time, professionally trained *Beamten*.[36] School attendance by students up to the age of eighteen was to be compulsory. "In principle," said the document, meaning for most children, attendance was to be absolved in a public elementary school (*Volksschule*) and a continuation vocational school.[37] But as there was the possibility for students to leave the *Volksschule* after it's first four years, called the *Grundschule* (Basic School), the authors spoke of an "organically developed" public school system which, extending upward from the common public *Grundschule*, offered as alternatives another four years of the elementary school or a transfer to one of several different middle or higher schools.[38]

When after completion of the four common *Grundschul* years for some ten-year-old students the possibility arose to attend a middle or higher school, they faced the issue of *Schulwahl*, the German term for school choice. Though the constitution did not use that term, it paid special attention to its implementation. It stated that the structure of the public school system corresponded to the diversity of occupational careers. A student's acceptance in one of the available schools therefore indicated his or her most likely later career path. The choice to be made, the constitution declared, was to reflect the child's "talents and inclinations; not the parents' economic or social position or their religious confession."[39] On parental desires other than those prompted by economic and social position or religious confession—the parents' understanding, for example, of the child's talents and inclinations, which might differ from that of the child's teachers—the constitution remained silent. The presumption, however, was that teachers, not parents, could best determine the child's talents and inclinations.

In response to demands by the churches and by the political parties that represented the churches' view, the authors of the constitution had granted parents the right to ask their local community for the opening of a public elementary school of their confession or ideology "as long as this will not interfere with the orderly operation of schooling." The parents' wishes, the constitution announced, were to be considered "whenever possible." Details were to be left to the laws of the *Länder*.[40] The authorization for parents to request confessional public schools could and was read by different parties in different ways. The majority of teachers favored public nondenominational schools (*Simultanschulen*) or, to a lesser degree, public secular schools. They argued that by giving special permission for the opening of public confessional schools the constitution implied that *Simultanschulen* were the "regular schools," *Regelschulen*, as the Germans called them. The teachers furthermore argued that they had earned *Regelschul* status for the *Simultanschulen* as a quid pro quo for having accepted the constitutional guarantee that, except in the secular public schools, religion was to be a recognized subject of instruction. The churches and their allies, however, maintained that the mandatory opening of public confessional schools upon parental request implied that public confessional schools had received the explicit sanction of the constitution and thus were to enjoy equal status with the *Regelschulen*.[41]

The constitutional clauses alluded to so far all referred to the system of public schooling. When it came to private schools, the constitution declared that those private schools that purported only to supplement existing public schools were to remain under existing regulations of the *Länder*. As supplementary schools could not bestow on their graduates vocational or professional entitlements, they were of little concern to the constitution's authors. However, those private schools that claimed to substitute for rather than supplement public schools required permission by the state and had to conform to the legislation of the various *Länder*. State authorization was to be granted as long as these schools compared favorably with public schools in their curricula, equipment, and the academic training of their teachers and as long as they did not promote the separation of students according to the economic position of their parents. State authorization was to be denied if the economic and legal position of the teachers could not be guaranteed.[42] Under these specifications there were to be, apart from legal ownership, scarcely any differences between private and public schools.

The constitution made clear that its clauses on private schools applied chiefly to post-elementary supplement and substitute schools. Having declared the *Grundschule* to be the public school common for all, the document implied, though it did not say so explicitly, that there could neither be private substitute nor private supplementary schools for the first four *Volksschule* years. However, the proponents of confessional schools ardently disputed that implication and, besides having succeeded in legitimizing public confessional elementary schools, they also achieved the insertion into the constitution of the phrase that private elementary schools were permitted, though only in communities in which public confessional or ideologically oriented elementary schools did not exist or where the school administration recognized a special educational purpose.[43] As table 3.6 shows, with slightly more than 96 percent of German public *Volksschulen* during the years of the Weimar Republic being confessional schools, the opening of private confessional elementary schools did not become an issue.[44] But the defenders of private preparatory schools (*Vorschulen*) fared less well. These schools or school classes drew children of well-off parents away from the public elementary schools in order to prepare them through an academic curriculum for entrance into a middle or higher school. In a brief, four-word sentence the constitution asked for their abolition.[45]

Table 3.6 Public Elementary Schools in Prussia, 1921–1932

Year	#/%	Protestant	Catholic	Jewish	Simultan	Total
1921/22	#	23,265	8,642	153	1,331	33,391
	%	69.67	25.88	0.46	3.99	100
1926/27	#	23,212	8,823	96	1,189	33,320
	%	69.66	26.48	0.29	3.57	100
1931/32	#	23,152	8,723	95	1,220	33,190
	%	69.76	26.28	0.29	3.68	100

Source: Based on Christoph Führ, *Zur Schulpolitik der Weimarer Republik*, 2nd ed. (Weinheim: Beltz, 1972), p. 344.

The National School Law of 1920 (*Reichsgrundschulgesetz*) then ordered all private as well as public preparatory schools to be closed as soon as possible. But state legislatures, beset in the following years by inflation and depression, found it impossible to accommodate students and teachers of these and other private schools, and the final disappearance of the preparatory schools did not occur until under the Nazis in 1936.[46]

Parents, the constitution declared, were to view the education of their children to bodily, spiritual, and social proficiency as their highest duty and natural right, and they were to exercise this parental duty and *Elternrecht* (parental right) under the watchful eye of the state.[47] The constitution decreed that parental opinion was to be considered as much as possible when parents asked for the opening of a confessional public school, or, where such a public confessional school did not exist, when they asked for the establishment of a private elementary school. But it gave no such guarantees when parents wished to have input into the administration of and the instruction in their children's schools, or when they wanted to register a complaint about a teacher or about disciplinary measures. As mentioned before, at the all-important occasion of *Schulwahl*, the constitution explicitly stated that parental economic, social, and religious position were to be disregarded. As far as the constitution was concerned, the protection of the *Elternrecht* pertained chiefly to enable parents to press for the opening of confessional schools.

This limited recognition of parental rights had been the result of fierce debates in the constitutional committee of the National

Assembly. In these debates, the viewpoints of government educational ministries, political parties, churches, and teacher groups clashed. They were eventually incorporated into the so-called Weimar compromise which, as already shown, endorsed the side-by-side existence of confessional and secular elementary schools together with the *Simultanschulen*, which were acknowledged to be the *Regelschulen*.[48] Throughout these struggles, the churches and political parties representing their interests made effective use of parents who favored the continued existence of confessional elementary schools and who wanted them to be recognized as *Regelschulen*.[49] They succeeded in safeguarding the parental right to confessional *Grundschulen* and relied on the *Elternrecht* to assure their continued existence and growth.[50] They argued that as long as attendance at a *Grundschule* was compulsory, it would be an indefensible coercion of conscience if such schools were state schools and did not permit parents a religious choice.[51] During the reaction that followed the revolution of 1918 they supported parental protests against the announced closure of the preparatory schools (*Vorschulen*) and the campaign of many parents to reduce the compulsory attendance of gifted children in the common public *Grundschule* from four years to three. While they could not prevent the eventual closing of the preparatory schools, they were successful in pushing through the adoption of the 1925 Elementary School Law, that permitted the admission of gifted children to higher schools after an attendance at the *Grundschule* of only three years.[52] In all these political battles the churches and their allies among the political parties demonstrated the effectiveness of their mobilization of middle-class parents.

But it had been the Social Democratic Party representing working-class families that had first demanded parental participation in the affairs of local schools. During the closing decades of the nineteenth century, party spokespersons pointed to precedents proposed during the Prussian reform era following the defeat of 1806. When Prussia's government then had set out to create a system of national education, it had appealed to the organs of civil society in cities and in the country to take on the local financial support and governance of their schools. As I described it in chapter 2, Johann Wilhelm von Süvern had argued in his never-adopted 1819 bill for a national school law that lay members, including "house fathers," be added to local school boards. After the failed revolution of 1848 the liberal pedagogue Adolph Diesterweg had similarly proposed that parents and taxpayers be included in local school societies. For both men, however, the leading idea had been that in a truly national system of education the taxpaying members of civil society were to serve as a counterweight to the state which was represented by administrators and clergymen. The laymen-and-women Süvern and Diesterweg had in mind were first of all citizen-taxpayers, and only secondarily were some of them parents. The Social Democrats' demand, however, encountered the determined resistance of the government which accused the party of intending to overthrow the existing political and social order. The party's members, the government argued, could therefore not be entrusted with participation in local school councils.[53]

Parent councils, however, became a reality as part of the revolutionary changes of November 1918. In Prussia and in Hamburg as well as in Saxony and Thuringia left-wing parties succeeded in having state governments mandate by law the establishment of parent councils in all public schools.[54] In Prussia the competencies of the councils were limited. Councils could not issue orders or regulations concerning the school's administration or instructional program, but they were authorized to give advice. They were neither allowed to visit classes and discuss complaints about teachers, nor could they call for meetings of parents without the consent of the faculty. Neither the school administration nor the teachers were bound to carry out parent council decisions. Opposition to the councils came from teachers who feared interference in what they considered their professional autonomy in the classroom. Administrators and teachers also remained suspicious that parents were going to be manipulated by clerical interest groups. When in the spring of 1920 parent council elections were held in Prussia, both left-and right-wing parties, the latter in close collaboration with the churches, put up candidates. Reflecting the political divisions of the country as a whole, the representatives of the conservative parties gained the upper hand over their antagonists on the left. Parent councils as an institution then began their work within the limited framework that had been assigned to them.[55]

While lay participation in school governance as envisaged in the nineteenth century by Süvern and Diesterweg had foreseen involvement of parents as taxpaying citizens and thus as representatives of civil society, the members of the parent councils of the Weimar Republic were elected as parents. Being laymen and laywomen, the members of these councils were to encounter continuing suspicions, regardless of their political, ideological, or confessional beliefs, and were never fully accepted by school administrators and teacher organizations as equals. The conservative parties and the churches viewed them as defenders of the *Elternrecht* to a public confessional school, and the left-wing parties intended them to represent and defend the interests of working-class families. But given their restricted assignments they could be effective neither in school policy and administrative matters nor could they interfere in curricular and instructional arrangements, which were considered the responsibility of the schools' instructional staff.

Discouraging as the overall picture of parental effectiveness may have appeared considering the widespread antipathy of teachers and administrators, in schools where progressive educators were at work, parental involvement in pedagogical reforms did in fact occur. In many cities and villages across the country parents supported teachers and administrators who were interested in reform-pedagogical experiments. They built on reforms that had been introduced during the closing decades of the nineteenth century as protest against the stultifying and bureaucratically run state schools in the major urban centers. They accepted the criticism of the heavy emphasis on academic learning voiced by Alfred Lichtwark, the Hamburg originator of the art education movement. They believed with him that German schools instructed their children but failed to educate them. They were inspired by such manifestos as Ellen Key's *Century of the Child of 1900*, Maria Montessori's *Pedagogical Anthropology* of 1910, and by the child psychology of Edouard Claparède. They looked for inspiration beyond the confines of the public schools toward such prewar pioneers of privately established country boarding schools as Hermann Lietz and his *Landerziehungsheim* at Ilsenburg, founded in 1898, Gustav Wyneken and his Free School Community Wickersdorf, opened in 1906, and Paul and Edith Geheeb and their Odenwaldschule which had begun its work in 1910.[56]

It is difficult to say precisely whether public or private schools deserve the greater credit for continuing the reform-pedagogical impulse during the 1920s. It seems clear that the original impetus was provided in the nineteenth century by the founders of private institutions. Based on the conviction that education should not be relegated to narrowly defined instructional institutions but must take place in a holistic environment that joined learning, art, and working, nature and culture, these school founders sought to avoid the institutional character of regular public schools. They believed that education must respect the child's individuality and be free to take its direction from the child's natural curiosity and interests. They eschewed set lesson plans and the uniform standards and regulations imposed by outside administrative authority. It is not surprising, then, that they could exist as role models and examples for the public schools only in some aspects of their instructional innovations. As institutions that prided themselves on their autonomous insularity they were not part of and could not serve as models for the public school establishment.

The *Landerziehungsheime* newly created during the years of the Weimar Republic—Bernhard Uffrecht's Free School and Work Community at Letzlingen in 1919, Kurt Hahn's Salem on Lake Constance in 1920, Martin Luserke's School by the Sea on the North Sea island of Juist in 1925, and a few others—held a special appeal for some sectors of Germany's economically well-off educated bourgeoisie. Parents whose children were repelled by the state-run schools with their heavy emphasis on subject matter and on conformity to social and curricular demands, found in these private progressive schools with their child-centered educational philosophies a welcome alternative. But they also had to recognize that this alternative came with a price. Not only were the expenses of tuition, board, and room considerable, but as the school incorporated their pupils into its community it effectively removed them from their families. Furthermore and ironically, as parents had taken the initiative in removing their child from a public school, they had handed him or her over to a boarding school's headmaster and staff. In these schools the voice of parents was strictly secondary to that of teachers and headmasters. Progressive education or reform pedagogy was an educators' domain.

Just as in the public schools the organized teachers saw to it that the sphere of influence of the parent councils was limited, so in these residential private schools the educational philosophy of the principal teacher and his colleagues was not to be challenged by parental interference.[57]

Parents were invited to more active participation in the nonresidential progressive private schools that flourished during the Weimar Republic. The Montessori schools took their guiding principles from Maria Montessori's child psychology and the pedagogy she practiced in her Casa dei Bambini, opened in 1907 in San Lorenzo near Rome. The *Freie Waldorfschulen*, opened first in September of 1919 in Stuttgart, sought to put into practice the anthroposophical philosophy of Rudolf Steiner. Echoing Wilhelm von Humboldt's early thoughts on education, Steiner taught that education should be free of governmental authority and release the creative potential of the pupil as a free, autonomous individual. He stressed the necessity for autonomy and self-direction of each school as a self-governing community of teachers, parents, and students. A Waldorf school was to be called into being by the initiative of parents. Its classroom practice was to be guided autonomously by the teacher. Education was to set free and was to give room to the creative energies of the student.

Another example of a reform-pedagogical private school that affirmed a close cooperation with its pupils' parents was the Home Teacher School founded by Berthold Otto in Berlin in 1906 and directed by him until his death in 1933. Otto agreed with Johann Friedrich Herbart, whose views I described briefly in chapter 1, that children learned best in small, uncoerced family circles. A home teacher or tutor was to guide the course of learning as had been common practice in bourgeois households of the seventeenth and eighteenth centuries. A conversational style was to characterize classroom proceedings which were prompted by the children's questions, not by a lesson plan. The children, not the school administration, set the disciplinary rules of everyday behavior.

During the years of the Weimar Republic reform-pedagogical initiatives also flourished in public as well as private schools. Because public schooling of necessity placed a high value on uniformity and equal treatment of the large numbers of pupils under its control, it could ill afford to adopt the private schools' attention to the children's individualized learning programs and the curricular and disciplinary flexibility this required. Other ways of countering the monotony of traditional instructional procedures and the boredom these created in the classroom had to be found. The so-called work school (*Arbeitsschule*), pioneered in the nineteenth century and, after the turn of the century, promoted by Georg Kerschensteiner, was one of the earliest attempts to introduce reform-pedagogical approaches into German public elementary schools. Convinced that active commitment to citizenship rather than passive knowledge about it should be the aim of citizenship education, Kerschensteiner introduced manual work and shop instruction in the Munich elementary schools. Such work experience in the *Arbeitsschule*, involving cooperation among groups of students, was a better preparation for the world of work than the passive absorption of knowledge in the *Lernschule*. The *Arbeitsschule* valued *können* over *kennen* (skills over knowledge). Because an emphasis on manual work rather than on academic learning suggested that the school aimed to prepare the students for limited career options, other reform pedagogues, above all Hugo Gaudig, defined the purpose of the *Arbeitsschule* as the encouragement of the student's sense of independence and initiative. This could be achieved, said Gaudig, the director of a higher girls' school in Leipzig, when students themselves determined the work to be done, the means and methods to accomplish it, and the criteria to evaluate it.[58]

Reform pedagogical endeavors in the public elementary school, however, could never completely disregard the conditions of schooling in a uniformly administered institutional setting. Teachers were state employees who had to fulfill prescribed educational tasks; students were neither self-selected nor, except for their religious affiliation, assigned to their school for other than residential reasons; the school community of teachers, administrators, students, and parents remained under the ultimate direction of the public whose mandates it could not ignore. These considerations persuaded Peter Petersen of the University of Jena during the late 1920s to propose what became the most successful comprehensive attempt at reform-pedagogical innovation in Germany's public schools. Petersen intended his Jena plan to be adoptable and adaptable to the widest variety of public schooling. The plan was a synthesis of many reform-pedagogical initiatives undertaken in Europe and the United States. It deviated from the usual reform-pedagogical emphasis on the individuality of the child and instead focused on the school as community. In this school community the students were to live under the educational guidance and direction of their teachers to be prepared for their eventual place in the larger community outside the school's walls. For Petersen, parents played an important role because in their daily contact with the students inside and outside the school they joined the life of the school with that of the surrounding community. Parental involvement in school projects, parental aid in decorating school rooms, and parental participation in parties and festivals all were means to serve the school's educational purpose of preparing children for life in community.[59]

The spread of reform-pedagogical information to Germany's public schools was aided by the German Central Institute for Education and Instruction. It was founded in 1915, and after 1918 was funded by all German states except Bavaria. While reform-pedagogical ideas were likely to find sympathetic listeners among a substantial number of teachers and school administrators, a special effort had to be made to convince and entice taxpayers and public school parents to support such reform efforts. As the country's central public collection agency of educational reports and statistics, the institute was quite successful in that endeavor. It made it possible that, according to Hanno Schmitt, by 1930 there were in existence—in addition to 21 private country boarding schools (*Landerziehungsheime*)—99 municipal experimental elementary schools, 17 municipal experimental secondary schools, and 62 public rural experimental schools. To these, Schmitt writes, ought to be added another nearly 300 schools that were not considered experimental schools as such but applied reform-pedagogical principles. All of them benefitted from the Institute's work and all of them, in one way or another, involved parents as well as teachers in their work.[60]

Parents seem to have played an even more prominent role in German public schooling as supporters of the rural school homes (*Schullandheime*) which increased in numbers between 1919 and

1933 from 20 to 255. These homes, located in rural areas, served as temporary places of instruction away from the regular schools in the cities. They were intended to provide for urban public elementary or secondary school children an experience of communal living in a healthy, stimulating environment. While they were not independent establishments privately founded and owned, they were not, except for minor contributions, funded by public agencies either. Parents, joined together in school societies, provided the mainstay of financial support for these homes.

Despite all these efforts, reform pedagogical initiatives did not bring forth the Copernican revolution in education that its promoters had hoped for. By the end of the Weimar period they still remained on the margins of the country's educational experience. In the public schools, administrators and teachers stayed at the center of the schools' daily routines. Parents played minor, supportive roles, sometimes invited, most often just tolerated. Only in some private reform-pedagogical ventures did students play a shaping role. Events occurring outside the schools—loss of the war, revolution, inflation, depression—did not favor the cause of educational reform. As Jürgen Oelkers pointed out, reform-pedagogical developments in Europe had "again and again been interrupted or deviated by political catastrophes. . . . Larger projects with verifiable long-term effects within the state sector of schooling never came about."[61] In the main, institutional rigidity, instructional tradition, and administrative bureaucracy prevailed over pedagogical reform.

In Hitler's Third Reich, then, the voices of an independent civil society were drowned out by the sound of marching feet and blaring horns of the party faithful. Hitler abolished the sovereignty of the *Länder* over educational policies. Directives and orders now emanated from Berlin and were transmitted and carried out by *Beamten* loyal to the National Socialist Party. Pedagogical reforms were stopped or continued in a minor key. Civil servants were told that to send their children to private schools was incompatible with their duties. Eventually, with few exceptions, private schools were closed by party orders. Some of the *Landerziehungsheime* survived by various strategies of adaptation. Others, such as Paul Geheeb's Odenwaldschule, were forced into exile, as was Kurt Hahn who continued his educational work at Gordonstoun in Scotland.[62] In the public sector, local school boards, parent councils, and school societies continued to function, but they did so under the tutelage of the Nazi party. The public experimental schools that had been established during the 1920s were closed as were the secular public schools.[63]

As it was the intent of the Nazi party to isolate and eventually eliminate Jewish citizens from German public life, the party decreed in 1933 that the number of Jewish students in German public schools could not exceed 5 percent of a school's enrollment or 1.5 percent of the annual admissions. Three years later Jewish children were segregated in separate classes of public schools until November 15, 1938, when they were forced to attend separate Jewish schools administered by the *Reichsvereinigung* of German Jews. These Jewish schools, including the residential country school home at Caputh near Berlin, were permitted to set their own curricula and valiantly strove to support their students in their daily struggle for survival in a hostile environment. While the country school home at Caputh was destroyed in 1938 following the *Kristallnacht* attacks on Jewish properties, the closure of the Jewish schools came in 1942 when the final deportation of Jewish adults and children to concentration camps had begun. Thereafter schooling for Jewish children took place in the concentration camps at Terezin, Auschwitz, and Buchenwald.[64]

In the months following the Nazi seizure of power in 1933, 46 or 33.5 percent of Prussia's 137 provincial superintendents and 115 or 22 percent of its 527 local superintendents were removed from their offices.[65] In all of Germany around 3,000 teachers were dismissed in 1933. Others, so a 1933 law stated, "whose prior political activities would not guarantee their unreserved commitment to the national state" could be given leave or be dismissed. Jewish teachers, unless they had been civil servants before the onset of World War I or had served as soldiers in the war, were forced into retirement. Whether dismissed or retired teachers were granted full, partial, or no pension was a matter of arbitrary administrative decision.[66] The party saw to it that school, home, and church as the traditional primary educational institutions were replaced by the party youth organizations with their programs of ideological indoctrination, premilitary physical education, and camp life. What the party liked to portray as an ostensibly enthusiastic independent youth movement was in reality a carefully disciplined state youth organization that served the party as a supportive counterweight to adult resistance. The catchphrase propagated by the Hitler-Youth—youth is to be led by youth—was adroitly taken over from the progressive youth movement of the 1920s and earlier. It artfully played the enthusiasm and idealism of youth against what the Nazi ideologues derisively called the stodginess of an older generation of parents and teachers. Unless they were faithful party members repeating the party's slogans, parents and teachers were thus relegated to the margins of the Nazis' educational world. Their opinions and wishes for their children and students counted for little. A party functionary's command would easily override a teachers recommendation or a parent's wish. Parental school choice had become a meaningless concept.

Chapter 4

School Choice in the United States after World War II

Racial Integration and School Choice

The decades of the twentieth-century's second half brought far-reaching transformations to the educational institutions of the United States. A people who had taken great pride in their reliance on local control over public education, who by the end of the nineteenth century had yielded only reluctantly to the growing influence of state administrators over their schools, had by the end of the twentieth century come to accept the imposition by the federal government of curricular standards and testing programs that would have appeared to earlier generations of Americans as complete repudiations of the American faith in the people's ability to run their own schools.

How did this remarkable change in traditional attitudes come about?

The role the federal government came to play in the nation's schools during the twentieth-century's second half had been prompted initially by concerns over the deep inequalities in school support across the country and by the continuing discrimination black American children experienced in their schools. School segregation along racial lines and poverty in rural and urban schools had led to several unsuccessful efforts to bring relief through congressional intervention. Attempts to pass bills for federal aid to education had failed again and again. It was not until the Supreme Court's unanimous decision in the *Brown* case of 1954 that the picture began to change.[1] The Court's statement that "in the field of public education the doctrine of 'separate but equal' has no place" and that "separate educational facilities are inherently unequal" convinced many Americans that the inequalities across the country in access to and in support of local schools demanded intervention by federal authorities, however time-honored they regarded the practice of local control.

The *Brown* decision also set into motion across the country a series of endeavors and proposals that ultimately looked toward parental choice as a means of either evading or of bringing about the Court-mandated integration in the nation's public schools. If parents were legally and financially enabled to send their children to the school of their choice, they could, if they wanted, avoid racially integrated and choose tuition-charging private schools. On the other hand, if through specialized curricula and equipment, integrated magnet schools could be made so attractive that parents would prefer to choose them, then choice could be made to work in support of integration, making unnecessary forced assignment and busing of students. And, quite apart from any debates about integration and segregation, choice appealed to parents who were dissatisfied with the bureaucracy of public schools and saw in school choice a free market alternative to what they considered a government controlled system.

Yet Southern opponents of the Supreme Court's *Brown* decision did not look to choice as their first means of circumventing the Court order.[2] They began their campaign of resistance by refusing outright to permit integrated schools and by closing the public schools. Only later did they introduce locally administered pupil assignment plans and "freedom of choice" policies, which were to enable white parents to send their children to white-only schools.[3]

The chief battlefields in these struggles were to be found in Virginia, Mississippi, South Carolina, Georgia, and Alabama. The Virginia State Board of Education reacted to the *Brown* decision by advising local school boards to continue segregation during the 1954/1955 session. In Mississippi voters adopted a constitutional amendment that "authorized the legislature to abolish the entire public school system, or to permit local school authorities to abolish parts of it." The legislature prohibited white students from attending public schools that admitted Negro students.[4] Later, in 1960, Mississippi amended its constitution to empower the legislature "at its discretion" to maintain and establish free public schools for all children. Similar actions were taken in South Carolina and Georgia. By August 1955 the Alabama legislature had passed a School Placement Law which, by omitting race from its pupil assignment criteria, was intended to preserve segregation without openly relying on racial selection. When the legislature next met in special session in January of 1956 it adopted a resolution that declared the *Brown* decision "null, void, and of no effect" and announced its willingness to resist its enforcement in Alabama with all means "honorably and constitutionally available to us."[5] It also passed bills in 1956 that called for amending the state constitution so that it would no longer recognize any right to education at public expense.

In Virginia, in the meanwhile, U.S. Senator Harry F. Byrd suggested in February of 1956 that "massive resistance" and "interposition" of state power, such as the Alabama legislature had set into motion, were "perfectly legal means" with which to oppose the Supreme Court's integration order. A similar sentiment was proclaimed in Washington by 19 of the 22 southern senators and 82 of the 106 congressmen who signed a "Southern Manifesto" in which they stated that they would "resist enforced integration by any lawful means."[6] The Virginia General Assembly, acting on these sentiments, cut off state funds from the public schools and closed any public school in which white and colored children were enrolled together. Massive resistance and interposition, however, proved to be short-lived. Virginia's cut-off of state funds and the closing order were invalidated as unconstitutional in January 1959 by both the State Supreme Court of Appeals and a federal district court. Later in the month the General Assembly abandoned "massive resistance" and the schools reopened their doors.[7]

What took the place of massive resistance and interposition were various measures that came to be known as state-encouraged "freedom of choice" programs. Alabama, Mississippi, and Louisiana had adopted such measures by authorizing voucher-like tax-rebate or tuition grants to families who preferred to send their children to segregated private schools. In Virginia a convention held in March of 1956 in Richmond amended the state's constitution "in such a way as to legalize tuition grants for education in 'public and nonsectarian private schools and institutions of learning.' "[8] The General Assembly quickly endorsed the move and enacted legislation to pay such tuition grants. Having also repealed the state's compulsory school attendance laws and having declared school attendance to be a matter of local option, freedom of choice programs now came to be introduced on local and county levels.

In Prince Edward County, where the supervisors refused to levy school taxes for 1959–1960 and closed the county's public schools, a private group, the Prince Edward School Foundation, began to operate private schools for white children. At first the foundation was financed through contributions. Later tuition was charged, but parents were reimbursed through state and local tuition grants. Citizens who contributed to the foundation were given real estate and personal property tax credits.[9] Prince Edward County blacks, refusing an offer by the foundation to set up private schools for them and preferring to legally press for integrated public schools, were without formal education for their children from 1959 to 1963. The Prince Edward County freedom of choice program, having been purposely devised to achieve the continuance of segregated schools, had effectively achieved that end.[10] A similar course of events unfolded in New Kent County, Virginia, where the County School Board maintained segregated schools until 1965 when it adopted a freedom of choice plan in order to remain eligible for federal funds. In Mississippi it was the White Citizens' Councils who in 1964 began their campaign for segregationist private academies that were to undercut both segregationist and integrationist public schools.[11]

When four states—Alabama, Louisiana, Mississippi, and Virginia—had enacted freedom of choice laws, the U.S. Supreme Court in 1967 in a *per curiam* opinion upheld an appeals court decision that had declared these state actions unconstitutional because "the United States Constitution does not permit the State to perform acts indirectly through private persons which it is forbidden to do directly."[12] And in a 1968 decision the Court stated that such "freedom of choice" plans might be acceptable when they "offer real promise of achieving a unitary, nonracial system" but they would not be acceptable when there are "reasonably available other ways, such as zoning . . ." As for the New Kent plan, the Court found that it had not dismantled segregation. "During the plan's three years of operation no white student has chosen to attend the all-Negro school, and although 115 Negro pupils enrolled in the formerly all-white school, 85 percent of the Negro students in the system still attend the all-Negro school." The plan, the Court stated, "has operated simply to burden students and their parents with a responsibility which Brown II [the 1955 *Brown* case decision] placed squarely on the School Board."[13]

Free Market Choice Plans and Vouchers

While the Supreme Court's *Brown* decision and subsequently the actions of the federal government enforcing it had spurred Southern state governments to propose their various freedom of choice plans, the next stimulus for such plans came one year after the Court's decision from a quite different source. In a 1955 paper the University of Chicago economist Milton Friedman offered a choice plan in which he did not refer to the use of "white flight academies" by Southern segregationists but inveighed against government financing and administering of public general and vocational education.[14] Such financing, he wrote, could be justified only by the "neighborhood effect" which shows that the education of children affects not only them and their families but their neighborhood as well. But government financing must not extend to the "actual administration of educational institutions by the government." Such a "nationalization of the education industry," he argued, leads to a government-run monopoly which through its uniformity and bureaucratic administration undercuts the public's investment in its schools and robs the schools of parental and community support. Vouchers permitting parental school choice, so his argument went, were the means by which a market system revitalized parental interest and commitment and improved the education of the country's children.

Friedman's 1955 proposal made the news just when the federal government's activities in education began to accelerate and, in turn, revived interest in progressive educational reforms. Civil rights leaders could not but see Friedman's reliance on markets as an unwelcome boost to Southern segregationists. They and their supporters vigorously condemned it and continued to urge the federal government to back integration in the public schools. Two years later the appearance in the skies of the Russian *Sputnik* brought more demands for federal support to public schools. It convinced the Congress to pass the National Defense Education Act of 1958 that appropriated federal funds for the support of science education in the nation's colleges and high schools. The increased emphasis that schools now devoted to academic work and competition brought forth a reaction among educators and students that produced a resurgence of unorthodox and progressive pedagogical ideas and practices and contributed to the student unrest of the 1960s. A veritable flood of pedagogical reform literature swept over the country. It encouraged the opening of alternative schools under both public or private auspices and brought forth calls for community control, homeschooling, and deschooling. The country's public schools came under heavy pressure from many sides.[15]

This interest in school reform also called attention back to Friedman's proposals for a market-based voucher system that he had published again in book form in 1962.[16] Later in the decade, his ideas were supported by liberal scholars like Theodore Sizer and Christopher Jencks who also spoke up for using vouchers to overcome the public school monopoly. Their main concern was to provide better education for the children in poverty-stricken inner cities. Sizer proposed a "poor children's Bill of Rights" which would empower the parents of poor children to choose among public and private schools and force these schools to compete for their students' voucher funds.[17] Jencks cited "inadequate public support and excessive bureaucratic timidity and defeatism" as causes of the educational malaise and the resulting social unrest in the urban ghettoes. He suggested that the big city public school monopoly for the poor be exposed to competition by private schools. To achieve that end he proposed tuition grants to poor children and management contracts to private organizations. To assure equity, Jencks stated that in an experimental voucher system that was to be funded by the federal

Office of Economic Opportunity, children from low-income households should receive larger vouchers than those from better-off families, and parents should not be able to supplement vouchers with funds of their own. The schools, on their part, would be required to abide by government rules of admission and to accept vouchers as full payment of tuition fees.[18]

With vouchers thus championed by both conservative and liberal scholars, the federal government stepped forward to fund an experimental program during the 1972–1973 school year near San Jose, California. As mentioned by Jencks in his proposal, the Office of Economic Opportunity (OEO) negotiated a compromise with the Alum Rock school district. The OEO had intended to test vouchers in both public and private schools but the Alum Rock district consented to the participation of public schools only.

The OEO agreed to supply the larger payments suggested by Jencks for poor families as compensatory vouchers for children who participated in the federal school lunch program. The district contributed the basic vouchers for all participants.[19]

The nature of the local opposition to the full-blown voucher program initially suggested by OEO becomes apparent in this quote from the Rand Corporation's report to the National Institute of Education:

> The school district's problem was how to accommodate OEO's desire for a voucher system without alienating its own constituencies—parents, principals, teachers, and central district staff. Parents wanted neighborhood schools. Teachers wanted the preservation of tenure and no "unprofessional" competition. Principals and central staff did not want to get involved in a popularity contest among public schools, let alone compete with private schools. Nobody in the school system wanted economic competition to dominate educators' behavior. And many Alum Rock parents, teachers, and principals were not yet ready to accept the voucher idea.[20]

The resulting compromise meant—at least during its first year—that far from testing a full-blown voucher experiment, the Alum Rock demonstration dealt with what has come to be called a "regulated voucher" system. It turned into a school choice experiment that focused on greater school autonomy and management responsibility. Choice entered into the picture as each participating school was divided into several minischools and parents and students thus could select different programs within their familiar neighborhood school. For teachers, the minischool programs also allowed greater initiative and discretion. If the Alum Rock experiment told little about the benefits or faults of vouchers, it did show that administrative decentralization and the minischool system increased parental involvement and the professional responsibility of teachers.

The Many Faces of School Choice

By the 1970s, then, school choice initiatives using regulated voucher systems had become a much discussed idea among liberals as well as conservatives. In California two legal scholars, John E. Coons and Stephen D. Sugarman, proposed such an approach in their 1978 book *Education by Choice: The Case for Family Control.*[21] Like Jencks and Sizer, they sought to minimize social inequalities and to maximize a family's ability to select for their children a public or private school of their choice. They had taken an earlier step by having prepared the argument for equalizing school financing in California and, in the *Serrano v. Priest* case,[22] had succeeded in lessening, though not completely eliminating, California's school funding inequalities. Their 1979 initiative campaign for regulated school choice, however, was less successful. As it included private schools, it failed to get on the ballot for lack of sufficient signatures. Still, voucher proposals of a liberal and socially conscious bent having achieved statewide recognition demonstrated that school choice as a reform movement was not necessarily the domain of segregationists and free market ideologues alone.

In fact, by the 1970s school choice measures of various kinds had come to be a winning proposition in many public school systems across the country in which they helped to break down rather than establish racial segregation. Disaffected students and their parents in many middle-class districts, seeking to revive a flagging interest in learning and to reach out across color lines, persuaded school boards to establish alternative schools that experimented with a variety of pedagogical approaches. Most of these relied on some form of progressive education as it had been known in the 1920s and 1930s. Many urban districts relied on magnet schools to encourage voluntary racial integration. These schools emphasized disciplinary curricular specializations and attracted students across the dividing lines drawn by segregated neighborhoods.[23]

Alternative and magnet schools, however, only served a selected few students and families who themselves had taken the initiative. More thorough-going choice reforms involving all students of a given school district or city took place during the 1970s and 1980s in New York City and Cambridge, Massachusetts. In New York City the reforms were to address the appalling inability of the public schools to educate the children of the poor and minority populations of the inner city. In Cambridge the purpose was to improve on a faltering effort to integrate the city's public schools.

The New York experiment followed in the wake of the disastrous teacher strikes and parent boycotts of 1968 that were occasioned by the struggle of the black citizens of the Ocean Hill-Brownsville neighborhood in Brooklyn for community control and decentralization of their public schools. The black parents' demands met with the determined resistance of the teachers' union, the United Federation of Teachers (UFT). The union, while advocating administrative decentralization, refused to grant powers over budget, curriculum, and personnel to parent-elected local school boards and preferred to retain them within the established bureaucracy. The confrontation ended in a stalemate and massive budget cuts and lay-offs of teachers.

It is against this background that in the spring of 1974, Anthony Alvarado, the superintendent in East Harlem's District 4, one of the city's poorest and most neglected districts, asked Deborah Meier to establish an elementary minischool in one wing of the district's public schools. He encouraged her and her colleagues to run their school on community- and child-centered progressive principles. The experiment succeeded so well that by 1994 there were fifty-two minischools, including one high school, which were housed in twenty buildings and run on progressive principles. As Meier explains it, small size and choice were essential to the schools' success.

> To make choice an effective force for change we need to provide incentives to districts to break up large schools and redesign them into many small schools easily accessible to families on the basis of choice. Small size is a major factor in improving schools and an absolutely essential one for the kind of pedagogical exploration we are talking about. Neither parents nor teachers can begin to talk together about what they want to do in schools where meetings take place in auditoriums and face-to-face conversation is a rarity.[24]

For Meier there was no doubt or hesitation: The principle of parental school choice was the key to a revitalization of the country's public schools.

Yet another tack was taken in Cambridge, Massachusetts. There, efforts to desegregate the city by mandatory assignment had allowed choice only to those parents who had selected for their children one of the city's magnet schools. Under the controlled choice plan first introduced in 1981 the elementary school district lines were abolished and all parents were free to choose among all of the city's public schools. To assist them in making well-informed choices, the city opened a Parent Information Center. With its help, 73 percent of the students who attended Cambridge schools between 1982 and 1986 could enroll in the school of their first choice, 18 percent in their second or third choice, and only 9 percent could not be so accommodated. Nonetheless, there remained popularity differences among schools along class and racial lines and between magnet and regular schools. To overcome them, the originators of the program stated, it should be the program's goal "to make all schools equally competitive in attracting students" and to educate parents "how to make choices and to demand changes in their schools . . ."[25]

By the 1980s the efforts on behalf of school choice and support for private schools had grown measurably. They were helped greatly by the wave of criticism that swept over the public schools in the wake of the 1983 report, *A Nation at Risk*. Employing an exaggerated rhetoric, the authors trumpeted, "if an unfriendly foreign power had attempted to impose on America the mediocre educational performance that exists today, we might well have viewed it as an act of war."[26] Their alarms were heard at the National Governors' Conference in 1986. The governors endorsed school choice as the best way to improve the nation's public schools. Governor Richard D. Lamm of Colorado, the chair of the task force on parent involvement and choice, stated: "If we implement broader choice plans, true choice among public schools, then we unlock the values of competition in the educational marketplace. Schools that compete for students, teachers, and dollars will, by virtue of the environment, make those changes that will allow them to succeed." But he shied away from a blanket endorsement of any and all choice plans and expressed his preference for government regulation. "Our recommendations are not for unrestrained choice," he wrote, "we suggest a state role in monitoring and limiting the use of choice, which will prevent the programs from having unintended consequences."[27]

While the governors were debating and preparing their report, a task force on teaching as a profession, established by the Carnegie Forum on Education and the Economy, countered *A Nation at Risk* with its report, *A Nation Prepared.* Declaring the raising of professional standards for the teaching profession to be their major aim, the members wrote that they did not believe "the educational system needs repairing; we believe it must be rebuilt to match the drastic change needed in our economy if we are to prepare our children for productive lives in the twenty-first century." Like the governors, they saw a market approach with magnet schools and vouchers and open enrollment among public schools within and across district lines as having considerable potential. They carefully avoided to mention private schools and they warned that such "approaches should be used only when there are policies in place to preclude racial or ethnic segregation, to ensure that there are no artificial barriers created that impeded any student's opportunity to attend the school of his or her choice, and to assure that students who do not exercise their option to leave their neighborhood school do not suffer as a result." Choice, they appeared to say, may hold many promises, but an unregulated market approach should be avoided.[28]

Charter Schools

A somewhat different note was struck in 1990 when John Chubb and Terry Moe published their *Politics, Markets, and America's Schools* and rejected most of the school reform efforts of the 1980s. "America's traditional institutions," they wrote, "cannot be relied on to solve the schools' bureaucracy problem . . ." Past reforms in the public schools, such as minimum competency tests for students and teachers, teacher empowerment and professionalism, school-based management, stricter accountability and bigger budgets, had been overly regulatory. Even such much praised innovations as those in Central Park, East Harlem, and in Cambridge, Massachusetts, did not satisfy them. They criticized the Central Park East Harlem choice schools for their being "subordinates in the hierarchy of democratic control, and what authority they have been privileged to exercise to this point has been delegated to them by their superiors—who have the right to take it back." As for the controlled choice project in Cambridge, they complained that parents and students were still part of a regulated public school system "firmly under the control of all the usual democratic institutions." Their bug-a-boo, they wrote over and over again, was the public school bureaucracy of "democratic control."[29]

To free public schools from this bureaucracy, Chubb and Moe promoted a system of publicly chartered choice schools, both public and private, in which authority had to be "vested directly in the schools, parents, and students. . . . As far as possible, all higher-level authority must be eliminated." The schools they envisaged were to be managed much like traditional private schools. They were to be financed through the students' scholarships or vouchers and could devise their own ways to handle their affairs. They would have to be allowed to fix the level of scholarships their students had to bring to be admitted and to set their own standards for admitting and expelling students. They were to be free to determine their forms of governance and regulations for teacher tenure.[30]

Yet despite their aversion to "democratic control" Chubb and Moe stayed shy of endorsing a completely free market system and despite their criticism of Cambridge's controlled choice they recommended their own version of it. This required wealthier districts to contribute larger amounts to the scholarship fund, restricted parents from supplementing their children's scholarships, allowed higher amounts for at-risk students, and permitted districts to tax themselves if they chose to increase their children's scholarship vouchers.[31] What they came up with was, in effect, a controlled free market.

The key to Chubb's and Moe's controlled free market was the public charter school. Charter schools, which began operating in 1992, broke radically with the concept of an expected uniformity of district school management and instead vested management authority in a group of community members, parents, teachers, and students. As Chubb and Moe had equated "democratic control" with a stifling bureaucratic uniformity, they saw in charter school independent management an opportunity for the expression of diverse teaching philosophies and cultural and social life styles, controlled only by whatever limits the chartering authorities had set.

The charter school movement spread rather quickly. As of April 2005, the Center for Education Reform (CER) listed 3,255 charter schools in operation in all of the states and the District of Columbia. California led the states with 533 charter schools, followed by Arizona with 509, Florida with 301, Ohio with 255, Texas with 234, Michigan with 216, and Wisconsin with 160. Charter schools obviously responded to a need unmet by most public schools. They paid particular attention to at-risk, minority and low-income students and to students who, as the editors of the CER report wrote, "are being failed by a 'one-size-fits-all' education system: Gifted and Talented students, teen parents, expelled and court-adjudicated youth, and non-English speaking children."[32] As reported in 2002, it is their average enrollment of 242 students—as contrasted with that of 539 in traditional public schools—that enables charter schools to perform their special services and attract and hold the loyalties of parents.

As charter schools are usually funded according to enrollment figures at the same rate that regular district schools are, yet are often confronted with start-up, building acquisition and maintenance costs as well as with the special demands of their students, they faced financial difficulties not encountered by regular district schools.[33] Another problem facing many charter schools was the question of authority and direction. It is comparatively easy to persuade parents dissatisfied with the services given to their children in regular public schools to support a charter school in which they, the students, and the teachers are to set the tone and manage policies and instruction. Yet to run a school on a democratic and cooperative basis is fraught with difficulties and has not always been successful.[34]

In Colorado the charter school idea of an independent school created at the initiative of parents, teachers, students, or by some other entity such as a university, a nonprofit, or for-profit company found a curious turn-around in the state legislature's Senate Bill 186 of 2000. The legislation, setting up a grading system for the state's public schools, provided that a school receiving an academic performance grade of "F" and unable to improve its performance within two years, shall be converted into a charter school. As the same legislation prescribed that a high-graded or improved school should be acknowledged and monetarily awarded, the mandated conversion of low-performing schools into charter schools was viewed by the legislature as a form of punishment for lagging academic performance. In Colorado, at least, charter schools may not always represent the voluntary search for academic improvement.

From the outset charter schools were intended to be part of the public school system though, as Chubb and Moe saw it, most of them were run like traditional private schools. To their detractors this anomalous relationship certified them as Trojan horses. By infiltrating a public school district they diminished its financial resources, subverted its contractual relationship with unionized teachers, and, by selecting their preferred students, evaded the public schools' responsibility to teach all children who live in the district. This has led some school boards to consider limiting the number of new charters they may be willing to grant.[35] As charters are also given to for-profit corporations—Education Management Organizations (EMOs) as they are called—the schools' public character came into question when public funds flowed into private hands.

On the whole, the charter school experience has been a mixed one. Parents in 2003, reported the *San Francisco Chronicle*, were just as desperate for educational alternatives as they were ten years ago. When they learned that eighty of California's established charter schools produced higher state test scores than traditional public schools, they remained eager to search for a charter school that their children could attend.[36] When they found a school like the Lionel Wilson College Preparatory Academy in Oakland they could feel gratified. Its cofounders were a former superintendent of the San Carlos Unified School District and the president of the California State Board of Education. Not only were these men experienced educators; but they also garnered considerable financial support from several philanthropic foundations and were able to open their brand-new school with the help of an $18 million private bond issue. They had every reason to envisage a successful future for their school.

But Lionel Wilson Academy was only one of thirteen charter schools in Oakland, and most of the others, said Oakland's superintendent, struggled on "nickel and dime" charters. Besides, California charter schools suffered from insufficient oversight that allowed operators to use one charter and open chains of schools in several school districts.[37] Reviewing reports of charter schools across the country, Howard Gardner wrote that while exceptional leaders may sustain charter schools for a time, they did not stay forever. He then added:

> In the absence of a group of first-rate teachers, a hefty endowment, or perhaps most important a sustained tradition of excellence supported fiscally and ideologically, the schools will eventually have a hard time competing with other schools, public or chartered. . . . It's not easy to build, finance, and run a school successfully. I also find it is difficult to conceive of a society in which there would be thousands of independent schools, each marching to its own tune. Any thought that so many institutions could be held accountable to reasonable standards is naive; and the potential for misleading advertising is appreciable.[38]

The diversity of educational programs and managerial arrangements—the very qualities that made charter schools so attractive to many parents—also made them vulnerable to pedagogical quackery and administrative malfeasance.

Private Schools

Parents who wished to evade altogether Chubb's and Moe's bug-a-boo, the public school bureaucracy of democratic control, could turn to private schools. Choice, not place of residence, governed their decision, although for members of religious congregations that choice may sometimes be limited. Private schools have been owned and governed by independent boards of trustees or by various religious organizations. Many require parental financial contributions in the form of tuition payments or fees, a requirement that also may inhibit choice. To what extent parents escape bureaucratic control and direction varies with the nature of the sponsoring organization. Whether their children will be exposed to a superior academic experience depends on circumstances that differ from school to school, though a desirable academic, spiritual, and social environment is often cited by parents as their reason for sending their children to a private school.

As a government publication reported in 2002, "on average, private schools have smaller enrollments, smaller average class sizes, and lower student/teacher ratios than public schools." With 77 percent, they show a somewhat higher proportion of white students than public schools do with 63 percent. They "are less likely than public schools to enroll limited-English proficient students or students who are eligible for the National School Lunch Program." Their most appealing claim is that their students "generally perform higher than their public school counterparts on standardized achievement tests" and their graduates "are more likely than their peers from public schools to have completed advanced-level courses in three academic subject areas."[39]

Even though enrollment statistics of private schools are incomplete, not always reliable, and are therefore to be used with caution, they allow insight into a few major developments. They show that during the half-century before 2000 the percentage of private school students—which had been on the increase during the century's first 50 years—now declined slightly from 11.4 to 9.9 percent. This decline was somewhat more rapid for high school students, from 10.5 to 8.6 percent, than for students in elementary schools where it moved from 11.7 to 10.5 percent (see table 4.1). At the century's end both the percentage and the enrollment figures for students in private elementary schools revealed a consistent decrease as the students moved from first to eighth grade. By the time they entered and moved through the private high school grades their numbers continued to decline. But, as indicated by percentage enrollment figures in private and public high schools, their rate of

Table 4.1 Enrollments in Thousands in U.S. Public and Private Schools from Kindergarten to Grade Twelve with Percentages of Enrollments in Private Institutions 1950–2000

Year	Enrollments			Percent in Private Schools	Enrollments in K–8th		Percent in Private K–8th	Enrollments in Grades 9–12		Percent in Private Grades 9–12
	All Schools	Public Schools	Private Schools		Public	Private		Public	Private	
1950	28,359	25,112	3,247	11.4	19,387	2,575	11.7	5,725	672	10.5
1960	41,762	36,087	5,675	13.6	27,602	4,286	13.4	8,485	1,035	10.9
1970	51,319	45,619	5,700	11.1	32,597	4,100	11.2	13,022	1,400	10.4
1980	46,208	40,878	5,331	11.5	27,647	3,992	12.6	13,231	1,339	9.2
1990	46,448	41,216	5,232	11.3	29,878	4,095	12.1	11,338	1,137	9.1
2000	52,012	46,856	5,162	9.9	33,487	3,908	10.5	13,369	1,253	8.6

Source: Data compiled from U.S. Bureau of the Census, *Historical Statistics of the United States, Colonial Times to 1970* (Washington: U.S. Department of Commerce, 1975), series H 412–432, U.S. Department of Education, National Center for Education Statistics, *Digest of Education Statistics, 2001*, Table 40, and Stephen Broughman and Lenore A. Colaciello, *Private School Universe Survey: 1999–2000*, Table 10.

retention was superior to that of their classmates in public high schools (see table 4.2).

At the outset of the new century private schools under religious sponsorship made up 78.4 percent of all private schools and accounted for 84.3 percent of all private school students. Among them the schools sponsored by the Catholic Church ranked first with 29.8 percent of all private schools and 48.6 percent of all private school students. Catholic schools generally were larger and attracted a greater diversity of minority students than other private schools.[40] Next in number of students were those schools classified as unspecified

Table 4.2 Number and Percentage Distribution of U.S. Public and Private School Students by Grade Level, 1999–2000

	Public		Private		
Grade Level	**Number**	**Percentage**	**Number**	**Percentage**	**Total Number**
Kindergarten	4,148,000	89.2	501,885	10.8	4,649,885
First	3,684,000	88.6	472,119	11.4	4,156,119
Second	3,655,000	89.1	449,093	10.9	4,104,093
Third	3,690,000	89.4	436,732	10.6	4,126,732
Fourth	3,686,000	89.7	425,140	10.3	4,111,140
Fifth	3,604,000	89.8	407,590	10.2	4,011,590
Sixth	3,564,000	89.8	403,114	10.2	3,967,114
Seventh	3,541,000	90.2	384,144	9.8	3,925,144
Eighth	3,497,000	90.4	369,579	9.6	3,866,579
Ungraded	418,000	87.5	59,446	12.5	477,446
Elementary Total	33,487,000	89.5	3,908,842	10.5	37,395,842
Ninth	3,935,000	92.1	336,224	7.9	4,271,224
Tenth	3,415,000	91.6	313,314	8.4	3,728,314
Eleventh	3,034,000	91.1	294,647	8.9	3,328,647
Twelfth	2,782,000	90.8	280,384	9.2	3,062,384
Ungraded	203,000	87.4	29,280	12.6	232,280
Secondary Total	13,369,000	91.4	1,253,849	8.6	14,622,849
Grand Total	46,856,000	90.1	5,162,691	9.9	52,018,691

Source: Based on Table 40 of U.S. Department of Education, National Center of Education Statistics, *Digest of Education Statistics 2001*, and on Table 10 of Broughman and Colaciello, *Private School Universe Survey, 1999–2000.*

Christian, followed by institutions of Baptist, Lutheran, Jewish, Episcopal, Assembly of God, and Seventh-Day Adventist sponsorship. Secular private schools and their students amounted to 21.6 percent of all private schools and 15.7 percent of all private school students (see table 4.3). They ranked relatively high among all

Table 4.3 Number of Private Schools and Students by Category, Fall 1999

School Sponsor	Schools	%	Students	%
Total	27,223	100	5,162,683	100
Catholic	8,102	29.8	2,511,040	48.6
Non-Catholic Religious	13,231	48.6	1,843,542	35.7
Amish	709	2.6	20,473	0.4
Assembly of God	486	1.8	75,255	1.5
Baptist	2,109	7.8	317,178	6.1
Brethren	54	0.2	8,328	0.2
Calvinist	150	0.6	40,802	0.8
Christian (unspecified)	3,611	13.3	533,008	10.3
Church of Christ	160	0.6	48,601	0.9
Church of God	145	0.5	15,140	0.3
Church of God in Christ	36	0.1	2,724	0.1
Episcopal	378	1.4	113,888	2.2
Friends	78	0.3	16,643	0.3
Greek Orthodox	28	0.1	4,614	0.1
Islamic	152	0.6	18,262	0.4
Jewish	691	2.5	169,751	3.3
Lutheran Church Missouri Synod	1,100	4.0	166,111	3.2
Evangelical Lutheran Church in America	121	0.4	18,400	0.4
Wisconsin Evangelical Lutheran Synod	359	1.3	33,815	0.7
Other Lutheran	70	0.3	4,369	0.1
Mennonite	414	1.5	24,262	0.5
Methodist	130	0.5	16,166	0.3
Pentecostal	472	1.7	33,201	0.6
Presbyterian	153	0.6	34,588	0.7
Seventh-Day Adventist	951	3.5	61,080	1.2
Other	674	2.5	66,885	1.3
Secular	5,890	21.6	808,101	15.7

Source: Stephen P. Broughman and Lenore A. Colaciello, *Private School Universe Survey: 1999–2000* (U.S. Department of Education: National Center for Education Statistics, August 2001), Table 2, p. 6.

Table 4.4 Number and Percentage of Students in U.S. Private Schools by Category and Grade Level, Fall 1999

Category	Kindergarten to Eighth Grade Elementary	%	High School Grades	%	Ungraded	%	Total
Non-Catholic Religious	1,456,963	79.0	374,204	20.3	12,413	0.7	1,843,580
Catholic	1,876,701	74.8	626,332	24.9	8,005	0.3	2,511,038
Secular	515,723	63.8	224,033	27.7	68,307	8.5	808,063
Total	3,849,387	74.5	1,224,569	23.7	88,725	1.7	5,162,681

Source: Stephen P. Broughman and Lenore A. Colaciello, *Private School Universe Survey: 1999–2000* (U.S. Department of Education: National Center for Education Statistics, August 2001), Table 10, p. 14.

private schools in their percentage of students in high school grades which, with almost 28 percent, exceeded Catholic schools with their 25 percent and non-Catholic religious schools with their 20 percent (see table 4.4).

Vouchers and the Supreme Court

The requirement of tuition and other fees prevented many parents, particularly those in the slums of inner cities, from enrolling their children in private schools. For them, vouchers offered an opportunity to overcome that obstacle. Vouchers could open the doors of private as well as public schools. In Milwaukee Polly Williams, a black assemblywoman, maintained that black children were inadequately served by magnet schools and busing which the city had used to break down racial segregation. She and her supporters wanted vouchers to be used for their children's attendance at charter schools. But they did not like the idea of a "controlled" free market. They argued that inner-city parents could on their own decide to which schools they would send their children. The Wisconsin legislature agreed and passed a Parental Choice Law that gave each of 1,000 low-income students a $2,500 voucher to be used for attendance at private nonsectarian schools. The schools had to agree not to charge fees that exceeded the voucher amount and to abide by federal antidiscrimination laws. After legislative and court battles, the program began in September of 1990. It proved to be so successful that five years later the legislature authorized the program to include as many as 15,000 students who were now allowed to attend religious as well as nonreligious private schools. A precedent was set for public vouchers to be spent in private as well as public schools.[41]

The inclusion of religious schools into the Milwaukee voucher program brought up anew a public debate over the separation of church and state. In decisions running into the second half of the twentieth century, the Supreme Court had not only upheld the First Amendment's ban on an establishment of religion, but had also rejected any attempt to read into the Bill of Rights "a philosophy of hostility to religion."[42] That same sentiment had been affirmed by the Court once more in the 1972 case of *Wisconsin v. Yoder*. The Court reasserted its finding in *Pierce v. Society of Sisters* that even though to provide for public schools was a state's "paramount responsibility," a state had to yield "to the right of parents to provide an equivalent education in a privately operated system."[43] The *Yoder* case dealt specifically with the contention of the Wisconsin Amish people that their children's compulsory attendance at a modern high school to age sixteen brought them "in sharp conflict with the fundamental mode of life mandated by the Amish religion." The Court agreed, and ruled that "the First and Fourteenth Amendment prevent the State from compelling respondents to cause their children to attend formal high school to age 16."[44] Instead, Amish parents were free to provide on their farms for their children's vocational education after they had completed elementary school.

A quite different version of parental choice gained national attention in 1978 when a Massachusetts court supported the home schooling movement in its struggle for acceptance as a legally permissible choice. This was not so much an example of "school" choice as of "education" choice, and for many parents it was based on their religious convictions. In *Perchemlides v. Frizzle*, presiding Judge Greany held that a parent's right "to chose alternative forms of education" was protected by the right to privacy which, in turn, grew "out of constitutional guarantees in addition to those contained in the First Amendment." That right, he said, applied to nonreligious as well as religious parents. The state could not deprive parents of this right arbitrarily, but only for serious educational reasons while observing all forms of due process.[45] The *Perchemlides* case, however, did not grant blanket permission for all kinds of home schooling everywhere, and the authority of states to impose reasonable regulations on home schooling had not been questioned.

To settle the vexing question whether state legislation in aid of church-related elementary and secondary schools violated the establishment and free exercise of religion clause of the First Amendment, the Supreme Court in 1971 in *Lemon v. Kurtzman* had consulted its own earlier decisions to establish a series of tests on which it might rely in the future. These tests demanded that the statute in question had to have "a secular legislative purpose;" that it had to show that the legislation's "principal or primary effect must be one that neither advances nor inhibits religion," and that it must not foster "an excessive government entanglement with religion."[46] The ruling became of major importance twelve years later when the Court upheld a Minnesota law that permitted state taxpayers to claim a deduction for expenses incurred for tuition, textbooks, and transportation for dependents attending elementary or secondary schools. Objections had been raised that the statute violated the establishment clause because parents of children attending parochial schools claimed the deduction. The Court, however, held that the law passed all three tests laid down in *Lemon v. Kurtzman*.[47] Parents who preferred to send their children to parochial schools could draw encouragement from that decision.

The series of Supreme Court decisions gradually weakened the First Amendment's prohibition against legislation that would establish a religion and gave greater prominence to the clause which banned legislation that would disallow religion's free exercise.[48] This trend was corroborated again in 2002 when on June 27 in *Zelman v. Simmons-Harris* the Court declared constitutional the use of vouchers to provide tuition aid to poor parents living in a failing public school system and sending their children to public or private religious or nonreligious schools of their choice. The long awaited decision appeared to settle the question whether and to which degree in states that enforce compulsory school attendance laws parents have a right to decide, or a voice in determining which school their children are to attend.

The establishment clause of the Constitution's First Amendment, which also applies to the states through the Fourteenth Amendment, raises the additional question whether parental choice, when it involves public funds flowing to religious institutions, violates what is commonly known as the separation of church and state. The Court's majority faced that issue directly and ruled that the voucher program did not violate the establishment clause. In the words of the chief justice, the program was religiously neutral and offered low-income parents genuine choice. Parental school choice and the use of vouchers appeared to have won a decisive legal and constitutional victory.[49]

Federal Initiatives

The Supreme Court's *Brown* decision of 1954 had called attention not only to the educational plight of the nation's minority populations but also to the educational neglect of poor children. When it became apparent that schools in urban areas with high concentrations of poor and minority populations were also schools of high student drop-out rates and low academic achievement, renewed efforts were made to pass a federal aid to education bill in Congress. President Johnson succeeded in this endeavor in April 1965 with the adoption of the Elementary and Secondary Education Act. Having promoted passage of the bill by emphasizing its purpose of providing financial assistance for programs "which contribute particularly to meeting the special educational needs of educationally deprived children," Johnson had avoided disputes over race and religion that had stymied earlier attempts to pass such legislation.[50]

The act broke new ground also in directing local educational agencies to assist poor children enrolled in private elementary and secondary schools with "special educational services and arrangements . . ."[51] These could include such materials as library books, textbooks, classroom supplies, and other instructional resources. In including this provision the Congress relied on the Supreme Court's decision in the *Everson* case of 1947, which had introduced the so-called child benefit theory. That theory held that, provided no account was taken of the children's religion and no funds were given to religious institutions, legislation which through a general program provided public funds to aid children in school did not breach the "wall between church and state."[52] As long as school districts obeyed these provisos, the Elementary and Secondary Education Act permitted public aid to children in parochial schools.

With the Elementary and Secondary Education Act the entry of the federal government into the field of American public education took on renewed force. Its twentieth-century precedents had been the Smith-Hughes Act of 1917 in support of vocational education and the National Defense Education Act of 1958. But the optimism and high expectations that accompanied the 1965 enactment proved unwarranted. The most disturbing news struck the public in 1983 with the *A Nation at Risk* report. It painted an alarming picture of America's deteriorating educational scene. In response the debate over education in Congress and elsewhere moved in two directions. One turned toward greater legislative support for school choice programs. Independent agents, both public and private, were to be encouraged to improve academic standards and outcomes through competition between private and public schools and within public education. The other pointed to standards-based reforms of public schools, accompanied by mandatory state and federal assessment and testing programs. In either case, political actors on state and federal levels were determined to impose their views on the country's elementary and secondary schools.

During the Reagan administration school choice under state supervision had received the endorsement of the National Governors' Conference in 1986. The Carnegie Forum of the same year had also considered it but refrained from mentioning private schools and warned against an unregulated market approach. The White House followed the same line. Speaking through Lauro F. Cavazos, the U.S. Secretary of Education, President George Herbert Walker Bush declared school choice to be "the cornerstone for restructuring America's system of elementary and secondary education." To signal the federal government's willingness to throw its support behind the school choice movement, Cavazos called for a White House Conference in January of 1989. Though participants remained divided on the question whether to include private and parochial schools in parental choice programs, they agreed that choice improved schools, empowered parents, and especially aided working poor and low-income families. They backed President Bush when he declared that "for this reason alone—for the benefit of empowerment it promises to our disadvantaged citizens—further expansion of public school choice is a national imperative."[53] President Bush did not mention private school choice.

The White House Conference of 1989 and the Charlottesville, Virginia, Education Summit of the same year, called by President Bush and attended by the nation's governors, shifted the focus from school choice to standard-based reforms. The conferences proposed six goals for the nation to achieve by 2000: school readiness for all children; school completion for at least 90 percent of all children; student competency in enumerated subject matters and citizenship; first-class world standing for the nation's students in mathematics and science education; adult literacy and lifelong learning opportunities for all Americans, and safe, disciplined and alcohol- and drug-free schools. The standard-based reforms, which were to accomplish these goals, were first defined by the National Council on Education Standards and Testing, created by Congress in 1991. They consisted of content standards for school subjects and performance standards for students. The standards were to be linked to assessment programs that tested students and measured a school's ability to deliver high-quality instruction with well-qualified teachers.

When in March 1994 the Clinton administration reauthorized the Elementary and Secondary Education Act of 1965, the goals proposed in 1989 were now incorporated and signed by President Clinton as the Goals 2000: Educate America Act. "By the year 2000, all children in America will start school ready to learn, . . . the high school graduation rate will increase to at least 90 percent, . . . [and] all students will leave grades 4, 8, and 12 having demonstrated competency over challenging subject matter including English, mathematics, science, foreign languages, civics and government, economics, art, history, and geography . . ."[54] The "Educate America Act" had added two more goals to the original six, teacher preparation and parental participation. States now were mandated to develop policies that would increase partnerships between schools and parents, and schools were to engage parents and families to support the academic work of children at home and to share decision making in the schools. In a further step to increase federal presence in public education, $10 million of federal funds were set aside to enable states to develop content and performance standards as well as tests that would measure student achievement.

The trend toward greater federal intervention increased markedly under President George W. Bush. In January 2001 President Bush announced his education reform proposals, which included school choice measures as well as standard-based reforms. He called for annual assessment tests in reading and mathematics in grades three to eight, rewards for high performing states, and sanctions for states "that fail to make adequate progress." There were to be start-up funding for public or private charter schools and a fund to "encourage innovative approaches that promote school choice." Special emphasis was to be given to reading programs, charter schools, and teacher training programs.[55] Although during his campaign for the presidency, then candidate Bush had championed vouchers, he did not mention them in his proposals. He did, however, tell reporters that parents whose children attended public schools that showed poor results for three consecutive years were to be given federal financial aid to help send their children to private schools. The money should come from Title I funds that otherwise would have supported the failing school. He did not think, he added, that the federal government "should try to impose a school voucher plan on states and local jurisdictions."[56]

One year later on January 8, 2002, President Bush signed into law the "No Child Left Behind Act of 2001."[57] Unprecedented federal requirements were to ensure that all American children will be enabled to improve their academic performance. The act greatly expanded standard-based reforms and accountability and encouraged parental school choice. It mandated the states to implement a single accountability system that would require annual testing at every grade level. As in the president's proposals of the preceding year, there was no mention of vouchers. The act, however, set aside roughly $200 million for charter schools to encourage school choice. Parents whose children attended persistently failing schools could now call upon their school districts to provide transportation for their children to a better public school or tutorial and other after-school services. While still shying away from outright support of private schools and voucher plans, the federal government now had assumed an unprecedented dominant role in governing America's schools.

Chapter 5

Schulwahl in the Post–World War II Period

Early Postwar Attempts at School Reform

When at the end of World War II the victorious allied powers published their plans for the reorganization of the German school systems, they accepted the traditional German view of state supervision over the public school system. They believed that through it they could best insure the realization of their declared chief aim, "the complete eradication of any and all traces of Nazism and militarism in Germany's public schools and the successful development of democratic ideas."[1] To this end they expected to establish through uniform state legislation the common school or *Einheitsschule* as the standard school type for all children.

The first to act on this program of reeducation were the Soviets who through their military administration in the Soviet Zone of Occupation (SBZ) caused German educators to publish a *Law for the Democratization of the German School* as early as May 31, 1946. Declaring that in the past "as a rule the doors of high schools and universities had been closed for the sons and daughters of the common people because their educational careers had been determined not by their abilities but by the financial status of their parents," they authorized a compulsory eight-year common school (*Grundschule*) for all children, to be followed by further schooling in a compulsory three-year vocational school (*Berufsschule*) or a four-year high school (*Oberschule*). The law did away with the traditional German bifurcation of schooling and *Schulwahl* at the age of nine or ten when those intended for academic and university preparatory training left the common school and transferred to an academic high school. As would soon become apparent, a student's admission to the *Oberschule* at age fourteen would depend primarily on his or her working-class background, membership in communist youth organizations, and nonparticipation in church activities. Students who did not meet these criteria and entered upon the three-year vocational school could obtain further advanced training in a two-year technical school (*Fachschule*).[2]

Next to propagate the *Einheitsschule* were the Americans. U.S. military authorities took as their model the American system of comprehensive public schooling for all students from the elementary through the middle or junior high school to the senior high school. They advised its representatives in the four *Länder* (States) of their Zone of Occupation in January of 1947 that all "two-pronged and overlapping tracks of schools should be eliminated. Elementary and high school were to represent two succeeding levels, not two different types or values of instruction."[3] Thus, like the Soviets, they eliminated any occasion for *Schulwahl.* Five months later the four occupying powers stipulated in a joint directive that in all parts of Germany the terms *Grundschule* and *Höhere Schule* were to describe two succeeding steps of schooling, not two overlapping basic forms of education.[4]

German educators reacted variously to these pronouncements, which sought to revamp their traditional two- or three-pronged school structure of *Volksschule*, middle school, and various high schools. While they were aware of the Soviet, American, and British origins of these proposals, they knew that the call for an *Einheitsschule*, a common elementary school for all students, also had roots in German soil. As I pointed out in chapter 4, during the Weimar Republic the German *Grundschule* law of 1920 had prohibited private elementary schools and had introduced a common four-year public *Grundschule* for all pupils. Socialist and other progressive school reformers, however, had fought for an eight-year common

Grundschule but had lost that battle and with it their major school reform proposal.[5] When communist schoolmen in the SBZ in 1946 introduced their *Law for the Democratization of the German School*, they highlighted that failed Weimar precedent because it allowed them to reject charges of trying to sovietize the schools. They endorsed the eight-year *Grundschule* with the subsequent educational paths of vocational, technical, and high schools as long fought-for democratic reforms of German ancestry. They also succeeded in 1948 to have their ideas incorporated in the school law for all four sectors of the city of Berlin, which was then under joint four-power administration. Here all children were to be taught in the eight-year *Grundschule* of a twelve-year *Einheitsschule*. Beginning in the ninth grade, those students who opted for academic schooling could enroll in the school's scientific branch whereas those who chose vocational training would enter in grade ten into a three-year vocational school.[6]

In the Western zones of occupation the *Einheitsschule* enjoyed initial popularity among reform-minded educators. Several of them in Hannover and Hesse welcomed the 1946 SBZ school law and in that year attended the first SBZ Pedagogical Congress in Leipzig.[7] In Schleswig-Holstein and the city-states of Bremen and Hamburg, too, school laws were passed in 1949 that established a six-year common *Grundschule* with succeeding three- or seven-year high schools of various specializations along practical and academic lines.[8] All of these laws did away with *Schulwahl* at age ten and postponed it to the age of twelve or fourteen.

Yet these early reform initiatives were not to last. Reminiscent of the early days of the Weimar Republic, the majority of educators, parents, and politicians in what had on May 23, 1949, become the Federal Republic elected to stay with or return to the familiar three-partite school structure that had been the rule then and throughout the Third Reich. During the years from 1950 to 1954 they saw to it that in the *Länder* that had adopted the six-year *Grundschule* these were reduced again to four years. After four-power cooperation broke down in Berlin during the blockade of 1948, the West Berlin city administration reconsidered and shortened its eight-year *Grundschule* to six years.[9]

These developments have led several commentators to speak of the restorative aspects of West German school politics during the early postwar years and of a missed opportunity for a thorough reform.[10] No doubt, a deep-seated resentment to what many saw as the imposition of a Soviet-style school system in the SBZ and subsequently the German Democratic Republic (DDR) and antipathy toward American reeducation efforts thought to be driven by misplaced missionary zeal fueled the rejection of the *Einheitsschule*. But more to the point was the revival after 1949 of political disputes of the 1920s and the insistence of partisans on all sides to remain loyal to their traditions and rebuild and strengthen a system that was familiar to them and most of their countrymen.[11] By the 1950s the Federal Republic's mood was set on reconstruction. Reeducation and reform no longer ranked high on the agenda.

School Policies in the SBZ and DDR

During the forty-year existence of the German Democratic Republic, a state-run and ideologically controlled *Einheitsschule* system was to persist though modified and adapted to the country's changing economic conditions. Some basic demands never changed. As the school law of 1946 made clear, the education of youth in schools was to remain exclusively under state control. Private schools were not permitted. Religion as a school subject of instruction was not offered. It was considered to be a task for the churches.[12] To ensure that a new, socialist spirit pervaded the schools, teachers were recruited preferentially from families of working-class background. By 1949, 70 percent of all public school teachers belonged to this "new class."[13] In a series of annual congresses between 1946 and 1949 the communist leadership of East Germany's educational establishment began to renounce its earlier reliance on German reform-pedagogical principles and replaced them with doctrines of Soviet educational science.[14] By 1951 the Central Committee of the Socialist Unity Party told the country's school teachers that, "based on marxist-leninist principles they had to transmit in their subjects of instruction the progressive results of science, especially of Soviet science."[15]

Nonetheless, while in their laws and administrative decrees the educational authorities of the DDR portrayed the eight-year *Grundschule* and the four-year *Oberschule* as a uniformly adopted *Einheitsschule* system, reality did not live up to this ideal. Uniformity existed on paper but could not be achieved in the thousands of *Grundschulen*, which included undifferentiated country schools, former urban *Volksschulen* with their structured organization, former middle schools as well as the lower and middle grades of former *Oberschulen* and *Gymnasia*. The *Oberschule* itself preserved the traditional differentiation of a modern language, a mathematics-natural science, and a classical language branch.[16]

The DDR's educational authorities themselves recognized these shortcomings and sought to address them in 1959 with their *Law for the Socialist Development of the Schools of the German Democratic Republic.*[17] The defining institution for the new school system was to be the ten-year polytechnical high school. It was to be the common school attended by all children. In the regime's marxist terminology this polytechnical school marked the transition from the antifascist-democratic to the socialist stage of national development. As the law said, the polytechnical school was to join schooling to production, mental to physical labor, theory to practice. In the lower grades manual education (*Werkunterricht*) was to be the chosen means for achieving this goal. Beginning in grade seven, vocational education (*Unterricht in der sozialistischen Produktion*) was to take its place. For the graduates of the ten-year high school further vocational training was to be had in two- or three-year vocational schools. In each of these, students should also have the opportunity to earn the traditional German high school leaving examination (*Abitur*), which opened the way to university attendance. A vocational and an academic education were to be had in the same school.

But despite the rhetoric stressing the ten-year high school as a common school offering general and work-related education for all students, the law also provided for a twelve-year polytechnical high school, the broadened high school (*Erweiterte Oberschule*) as it was called. In grade nine its students entered upon a university preparatory course in one of three electives: the natural sciences, modern, and classical languages. Here too, special emphasis was to be given to vocational or scientific preparation. Between graduation from the twelve-year high school and entrance into a university, the students then had to absolve a one-year vocational internship.

There was inherent in this legislation an obvious discrepancy between the common ten-year polytechnical education compulsory for all and the twelve-year polytechnical path with academic electives for a chosen few. It points out the conflict between the ideologically mandatory concern to obtain for the country's producing workers and farmers the advanced education which had been denied them in the past and the country's need for academically and scientifically highly trained specialists. It is all the more ironic to note that for all intents and purposes the three-pronged *Erweiterte Oberschule* returned to the traditional German gymnasial way through the *Abitur* to the university. For the educators debating the "correct" meaning of a polytechnical education, the discrepancy pointed to here was to present an ever-recurring dilemma.[18]

One of the problems bedeviling polytechnical education was that in all too many cases it had been turned into practical vocational training and had neglected its original more general scientific and prevocational preparatory intent. It became necessary for the educators of the DDR to remedy this deficiency. "Mastering the technological revolution and developing a socialist community," they stated in the February 25, 1965, *Law for the Unified Socialist System of Education*, "now required a higher quality of the unified socialist system of education."[19] The polytechnical *Oberschule* had to divest itself of its vocational training, concentrate on its prevocational introduction to the world of work, and combine it with a more traditionally academic scientific education. Beginning with the academic year 1967/1968 the *Oberschulen* were to inaugurate special preparatory classes in grades nine and ten that, for all practical purposes, were to adopt the curriculum of the *Erweiterte Oberschule* or, in yet another variation, were to emphasize instruction in modern languages. The *Erweiterte Oberschule* itself was shorn of its three different tracks and, beginning in 1969, was to offer elective classes in mathematics and the natural sciences, the social sciences, modern languages, and in music and the arts.[20]

The selection of pupils for the preparatory classes was to be carried out by a county school board in consultation with the principals of the schools involved. While among the criteria for admission the pupils' academic record was listed first, the politically charged demands for good behavior, loyalty to the state, and extracurricular activities followed, as well as a reminder to the board that when students later moved on to the eleventh grade of the *Erweiterte Oberschule*, the class-related social composition of that grade would have to correspond to the social class structure of the general population.[21] For the admission to the *Erweiterte Oberschule* itself the county school board had to appoint a special commission which was to include, besides members of the board and the schools' principals, representatives of the Free German Youth and of the trade unions. The commissions were charged to keep in mind that the children of parents who were especially active in building socialism were to be given preference; so, too, were the children of working-class parents, especially of those employed in production, and of members of the agricultural production collectives.[22]

The new dispensation tried to bring together the requirements of a unitary system—no parallel grades nine and ten on the high school level—and the emphasis on specialized education. Ever since the 1950s, special schools had been opened in music, theater, dance, the fine arts, and sport. Special schools and classes in mathematics, the natural sciences, and in technical fields, beginning usually with grade seven, were introduced during the 1960s.[23] By the 1970s the new special schools came to encompass grades nine to twelve and, like the *Erweiterte Oberschule*, were meant to prepare their students for the *Abitur* and entrance into a university. The need for specialized experts now overrode the earlier intent to create an all-encompassing *Einheitsschule* and instead created the insistence on a unitary system. In it, as the country's constitution of 1968 declared, the criteria for a student's advancement were "academic achievement, social needs and a consideration of the population's social structure."[24] One of the by-products of the keenly felt impact of the "scientific-technical revolution" was the fact that, on paper at least, academic achievement was to outrank the social criteria of a student's working-class background, his or her participation in communist youth organizations, and his or her abstaining from church-sponsored activities. Nevertheless, as Oskar Anweiler remarked, "the granting or withholding of educational opportunities became a not to be underestimated instrument of stabilizing the political system."[25]

One characteristic aspect of the SED's program for the country's youth was the intimate coordination of physical education with political indoctrination and preparation for service in the armed forces. In July of 1968 the Ministry of Education in conjunction with the Free German Youth and the Society for Sport and Technic issued guidelines for a competition to be held in grades eight to ten of the polytechnical high schools.[26] The event was to begin with lectures on the East German People's Army and the alliance of the Warsaw Treaty. The athletic competitions for boys and girls were to include target practice with both BB and small-caliber guns and hand grenades as well as war games inspired by "goal-oriented political motivation."[27] By February of 1973 the Society for Sport and Technic and the DDR's German Red Cross were officially asked to take over premilitary and medical training in the *Erweiterte Oberschule* and other specialized high schools in order, as the party declared, "to further the readiness and the ability of all citizens to defend our socialist state."[28]

Except for the introduction of compulsory military training in the schools of the DDR toward the end of 1978 and, in 1980, the abolition of preparatory classes and the reduction of the *Erweiterte Oberschule* to grades eleven and twelve, no significant changes occurred during the 1970s and 1980s. The success of the polytechnical high school as well as the relative numerical importance of its twelve-year university preparatory branch may be best judged by a few statistics. Despite the legal requirement of compulsory attendance, during the 1980s roughly 7 percent of the high school's students dropped out after the eighth grade, and a little over 86 percent of the 16- to 17-year age group graduated after completing all 10 grades. For the *Erweiterte Oberschule* the percentage of the country's 18- to 19- year age group graduating in 1984 from the twelfth grade with the *Abitur* came to 8.4 percent. If one adds to these the roughly 4 percent of students who earned their *Abitur* in one of the institutions of vocational and technical education, the percentage of students earning the right to university attendance came to between 11 and 13 percent. This is not a very impressive figure, and it explains to some extent the regime's emphasis on the responsibility of the schools for the "scientific-technical revolution" that was to overcome the country's economic difficulties.[29]

Given the regime's determination to prescribe and supervise the country's educational policies and to keep all school activities within the confines of its unified socialist education system, it is pointless to raise the topic of school choice.[30] The concept of the *Einheitsschule* and the system's structure did away at the outset with the German practice of *Schulwahl.* The decisions for choice that remained—whether a youngster was to be prepared for a vocation or for a professional career—were made by teachers and school administrators in consultation with representatives of the Free German Youth and of industries sponsoring vocational training. They were guided for the most part by social and political considerations and only during the 1970s and 1980s did educational criteria become more important.

The 1949 Constitution of the DDR stated that parents were to participate in the school education of their children through advisory boards elected by parent assemblies at the beginning of each school year. The boards were to support principal and teachers and to enlighten parents pedagogically to assure cooperation between education in school and homes. In addition to the elected parents, board meetings were to be attended by the school's principal, the head of the school's youth organization, a representative of the women's league, and the head of the teachers union. The labor union of a sponsoring industrial plant was also asked to delegate a standing representative. Meetings of the board were to take place as needed, but at least three times a year. These governmental instructions were issued in April of 1951, reaffirmed in October of 1955 when the meetings of the board were increased to once a month,[31] and issued anew in January of 1960. At that time, each parent advisory board was to be assisted with several specialized commissions whose members were to be recommended by the school administration, political organizations, and the sponsoring industrial plant. In addition there were to be parent collectives for each of the school's classes.[32]

Apparently, even this politically weighted system did not always produce the results desired by the SED. On August 9, 1967, the Central Committee of the Socialist Unity Party issued a directive that exhorted "all leading party organs to heighten the common responsibility of school, parents, youth organizations and sponsoring industrial and commercial plants to coordinate their efforts . . . for a continued realization of the law for the unified socialist system of education." The directive appealed especially to the party members among parents and urged them to spur on their children "to obey the rules and norms of socialist morals and ethics."[33] As if this were not enough, a decree of January 15, 1970, ordered the creation of special commissions to supervise the election of parent advisory boards. These commissions were to include the school's principal, three to five parents, a representative of the Young Pioneer organization and of the city or village council. The National Front, the Women's League, the sponsoring plant, the school's trade union, and the Free German Youth organization all were entitled to membership on the commission.[34]

It is clear from the description of the boards' functions and the inclusion of nonschool representatives that the boards did not exist to encourage parents to bring to the attention of the school authorities concerns they might have had about their children's progress in class or to voice discontent and complaints about school policies. They were meant instead to insure political conformity, to aid the authorities in enforcing discipline through all aspects of the students' lives, and to "enlighten" parents about the school's decisions for their children's future.

Schooling in a Reunified Germany

With the collapse of the communist regime of the DDR in November of 1989, the signing of the German Unity Treaty in August of 1990, and the final session of the Joint Educational Commission of the two German states one month later, the way had been prepared for the reorganization of the educational system of the former DDR—now called the new *Länder*—and its incorporation into that of the Federal Republic. The *Einheitsschule* and the unitary socialist educational system were gone. Public education in the unified Germany was organized in the traditional German dual system of academic and vocational education and its three-pronged arrangement of elementary, middle, and higher schools with its *Schulwahl.* Children would now begin their schooling in the *Grundschule*, the first four years of the *Volksschule*, and, unless at *Schulwahl* they changed over to a six-year *Realschule* or a nine-year *Gymnasium*, they would proceed to the upper four or five years of the *Volksschule* commonly called *Hauptschule.*[35] In many states grades five and six of the *Hauptschule* offered an orientation level called the *Förderstufe* which would allow a student to change-over to a *Realschule* or a *Gymnasium* at the completion of grade six. Some *Länder* had introduced a comprehensive school called the *Integrierte Gesamtschule* which usually comprised grades five to ten. Whether a state introduced the *Gesamtschule* at all, how many or how few of them it permitted, and whether it declared these schools to be regular (*Regelschulen*), optional (*Angebotsschulen*), or experimental schools (*Versuchsschulen*), depended on the political party in power. States like North Rhine-Westphalia and Hesse, dominated in 1987 by the Social Democratic Party, opened 95 and 71 *Gesamtschulen* respectively, whereas CDU/CSU states like Baden-Württemberg, Bavaria, and Schleswig-Holstein permitted only two each.[36]

Educational policies in both the old and the new *Länder* are governed by the *Grundgesetz*' (Basic Law or Constitution) fundamental dictum which states that "the entire school system stands under the supervision of the state." While this applies to both public and private schools, the state's supervisory power has always been and still is felt with particular immediacy in public schools. When school personnel oppose the wishes of parents or students or reject their requests, they act with the full authority of the state. When parents ask for reforms or changes in policy their requests are often seen as challenges to an entrenched bureaucracy with its innumerable edicts and regulations, well established in law and custom. To bring about change in such a system is an intimidating task.

The general restorative character of school policies pursued in the Federal Republic after 1945 also discouraged a resurgence of teacher-led pedagogical innovations such as had characterized the years of the Weimar Republic.[37] Incentives for change or reform were likely to succeed only when they originated with or had been approved by state authority. Examples of this are the measures taken to make less impermeable the walls of separation among school types and the laws amended by some state ministries of education to allow for greater school autonomy in arranging organization, curricula, and resources. They allowed principals to respond more freely to requests of parents, teachers, and students. In many secondary schools they opened alternative paths of study for students to select and pursue. Though, as one commentator remarked, these changes have not been very spectacular, in the German context of statist traditions they have had "the quality of a shift in paradigm."[38]

The *Grundgesetz*, which now applied to all German states, decreed that religious instruction was a regular school subject to be taught in agreement with the principles of the religious communities. Parents were free to decide on their children's participation in that subject. Similar to the constitution of the Weimar Republic, the *Grundgesetz* also guaranteed individuals the freedom to found, develop, and direct secular and confessional private schools on the condition that such schools are subject to the laws of the individual *Länder* and that teacher preparation and curricular aims and equipment are equal to those of the public schools.[39] In the selection of their students private school masters may not consider parental income, and they must guarantee the economic and legal status of their teachers. Private elementary schools, however, may be licensed only when state authorities recognize a special pedagogical interest or when their founders intend to open a common, confessional, or ideological school that does not already exist among the local public schools.[40]

State laws further define the tasks of German private schools as "supplementing and enriching the public school system and promoting it through special forms of instruction or education."[41] Only when in the judgment of public officials a private school complies with all the constitutional requirements, will the government recognize it as a substitute school (*Ersatzschule*). Such a school then can offer its graduates the same entitlements as the public schools obtain for theirs. If, however, it does not meet these conditions, it can be classified only as a supplementary school (*Ergänzungsschule*) whose graduates do not earn the entitlements of public school graduates. This requirement may and sometimes does force a private school to deviate from its own pedagogical philosophy and practices in order to obtain entitlement privileges for its graduates. There is no legal way in which German private schools can extricate themselves from state involvement and supervision which extends also over the moral conduct and political loyalty of their teachers. German private schools have in effect become quasi-public institutions.[42]

Without outside financial aid few German private substitute schools are in a position to abide by the requirements of the Basic Law not to discriminate among their students on the basis of parental income and to provide adequate financial compensation and legal protection for their teachers. Not being able to meet these requirements they cannot then obtain the state license as a substitute school. Acknowledging this predicament, German courts concluded that the Basic Law's guarantee for the founding of private schools implied a state obligation to provide financial subsidies to substitute schools that required them. The amount and kind of subsidies, however, differ in the various states. To make matters worse, by the spring of 2003 the waiting time between the opening of a new private school and the receipt of state subsidies stretched from three to five years, preventing many planned schools from going ahead.[43] State subsidies very often are insufficient to meet a school's costs, and parents are asked to cover the deficit. According to a decision by the Federal Administrative Court parents cannot then claim reimbursement from the state.[44] Parental choice thus encounters limits that are particularly severely felt among low-income families.[45]

"There is no question," declares a report issued by the Max-Planck-Institute for Education Research, "*the* school of the Federal Republic is the state [or public] school."[46] While in other European countries the percentage of private schools hovers around 20 percent, in Germany it stagnates below 6. But there is also no question that during the 1990s German private schools, both substitute and supplementary, enjoyed renewed popularity. According to the Statistische Bundesamt, the number of students in private schools rose from 570,000 in 1992 to 740,000 in 2000. The share of private schools among vocational schools stood as high as 34.6 percent, and among private schools of general education the *Gymnasia* showed the highest percentage with 11.6 percent. Kathrin Spoerr reported in *Die Welt* of April 7, 2003, that up to 20 percent of parents inquiring about admission for their children in private schools were willing to enroll them even if they had to pay tuition, while 30 percent would do so if attendance were free.

What brought about the renewed interest in private schools? Several reasons are usually listed: The public schools' inflexibility of curricular, timetable, and instructional arrangements and traditions; the bureaucratic facelessness and impenetrability of public school administrative offices; the unwillingness of public school authorities to listen to the concerns of parents and students; the often inadequate academic quality of instruction as revealed by standardized tests; and the resistance to offer instruction and programs to students throughout the day—a complaint voiced most often by working parents and highlighted by the fact that, in international comparisons, Germany ranks last in this respect. German public schools and their administrators are perceived by many parents as part of a sterile state government, limited in their ability to act autonomously, to respond flexibly to local conditions, or to explore cooperative arrangements with youth clubs, parental groups, and service organizations. All these liabilities work together to persuade increasingly more parents to consider sending their children to a private school.[47]

Among Germany's private schools, Catholic and Protestant confessional schools rank highest in numbers. They enroll roughly three-quarters of all private school students, and two-thirds of them attend Catholic schools.[48] Their numerical increase during the past few decades can be attributed in part to the secularization of German schooling during the 1960s when many public schools lost their confessional character. As a result both churches deliberately fostered confessional alternatives to the secular public schools. Catholic elementary schools operate on the basis of an assumed mutual agreement among pupils, parents, and teachers on the aims of education. Based on their understanding of Christian schooling as furthering a balanced education of the whole child, these schools seek to soften society's influence on the growing child, the early exposure to commercialism, and the early introduction to educational specialization. During the early 1970s, the numbers of these schools increased and attracted 9 percent of their pupils from non-Catholic families. Catholic university preparatory schools, higher schools for girls, as well as vocational and special schools likewise were successful in attracting larger numbers of students. Given the different purposes of these schools, parental motivations varied, ranging from disapproval of coeducation to a desire for a religious education and to preference for a full-day comprehensive school. For most of these schools parental participation in school events has been an important part of school policy.[49] Much the same can be said of Protestant parochial schools which also strive to create religious communities of pupils, parents, and teachers in which all share in the running of school affairs. A special interest of Protestant schools has always been and remains a close cooperation with religious social work agencies.[50]

Among secular private schools—known officially as schools of independent sponsorship (*Schulen in freier Trägerschaft*)—those intended to pay special attention to children with learning difficulties may be distinguished from those that offer pedagogical alternatives. Among the latter we may count the country boarding schools (*Landerziehungsheime*), which emphasize for their students individual development and social engagement, outdoor exercises, sea rescue service and firefighting, social work and manual education. Their students prepare for academic work and receive training in technical and other manual occupations. Then there are the *Waldorfschulen* which base their approach on the pedagogy of Rudolf Steiner. They emphasize the arts, dance, and music and reject the conventional ways of grading and student selection. Others include the *Montessorischulen* and the independent schools organized in the Federal Association of German Private Schools (VDP) as well as specialized private schools, such as the bilingual *Kennedyschule* in Berlin. Finally, there are commercial schools and private teachers who sell their wares on the market place and offer additional tutoring for students who desire their services.[51]

Schulwahl, Social Class, and Ethnicity

Among all the forms of school choice, German *Schulwahl* belongs in a category all by itself. It is intimately bound up with the German three-pronged school system and refers to a selection process rather than a true choice, an *Auswahl* rather than a *Wahl*. It is conducted and supervised by teachers and other school authorities. The *Grundgesetz* declares that parents have a right to determine their child's education, but the state, represented by school authorities, also has a right and a duty to act in the child's best interest. Thus, under the guidance of school authorities, parents are asked to indicate whether they want their nine- or ten-year-old child to advance to the next grade in the common school, the *Hauptschule*, or whether they want him or her to transfer to an advanced school, the *Realschule* or the *Gymnasium*. Parents know that this decision is of crucial importance for their child's life because it will determine not only their child's future schooling, but most likely also his or her occupation or career. Thus for parents *Schulwahl* brings heightened anxiety and worry.

Though the regulations that prescribe the methods by which *Schulwahl* is carried out vary in detail from *Land* to *Land*, they are similar in their overall intent. They follow in the main the recommendations issued in 1999 by the Standing Conference of the Ministers of Education of the German *Länder*.[52] Parents are encouraged to take the first step and make their wishes known concerning the desired school for their child. Then teachers and other school authorities will consider the child's past academic record, the results of an examination given at the time, and, based on this and similar evidence, judge the child's ability to meet the demands of the school the parents had chosen. The recommendations state that the transferring and receiving schools must cooperate in their judgment through mutual visits of teachers and students, joint conferences, and special teacher preparation sessions. Both schools are to advise parents "intensively and continuously," and the vote of the transferring school has to be issued together with detailed advice to the parents.[53] The recommendations also make it clear that space limitations and concerns for the maintenance of elevated academic standards in the advanced schools require that *Schulwahl* remain a process of selection in the hand of school authorities. It is hoped nonetheless that parents will find the recommendations of the teaching staff persuasive and accept them.

When parents disagree with the decisions of school authorities, the laws, regulations, and recommendations of the different *Länder* come into play.[54] In Berlin, for example, parents or guardians are asked to consider the transferring school's recommendation and to decide on the basis of information provided by the school about available options. Should they reject these recommendation, they are to be given another opportunity to receive advice. Because of possible overcrowding in the selected school, they are asked to indicate second and third preferences. If their decision has been for a *Realschule* or *Gymnasium*, their child will be admitted on probation for a half a year. Should it become evident that the student does not meet the school's standards, he or she will be asked to change schools.[55] In Brandenburg the regulations say that "abilities and interests of the child as well as the will of the parents determine school career." The greatest weight, however, is to be given to the transferring school's certificate of abilities, achievements, and interests. If there is an oversupply of applicants, the principal of the accepting school will select among them, basing his or her decision on the transferring school's certificate and on the applicants' fitness for the accepting school. Parents have the right to request a hearing, but nowhere are they assured that their wishes will prevail.[56]

These laws and regulations try hard to downplay the incidence of disagreements between parents and school authorities. In the State of Hesse they declare bluntly that disagreement and conflict are exceptions. "In most cases," they state, "an open, trusting talk with teachers will help."[57] When disagreements threaten, school authorities don't necessarily review parental objections. They seek to defend their advice or decision by arguing that they have sought an adequate balance between the "objective" criteria of test results and the "subjective" recommendations of the teaching staff. While they readily concede that they are not infallible, they argue steadfastly that their judgment has been reliable for three-quarters or two-thirds of their students. This, they argue, is as much as can be reasonably expected. In the last analysis parents are asked to trust and accept the school's recommendation.[58]

Schulwahl as a practice is integrally embedded in the long-established three-partite German public school system. As long as the divisions among the various school types exist, *Schulwahl* will continue to segregate students and their schools by academic criteria.

After the end of World War II some of the West German *Länder* had extended the length of common schooling from four to six or eight years and these changes, where made, modified *Schulwahl* to some extent. The eight- or ten-year *Einheitsschule* in the DDR even managed to abolish *Schulwahl* altogether. But in today's unified Germany few voices are heard that support modifications or abolition. *Schulwahl* is not likely to disappear.

During the 1980s observers began to note a marked lessening of the social class divisions that had been inherent in the *Schulwahl* system. They referred to the greater permeability of school structures as evidenced in the *Förderstufe* of the *Hauptschule* and to the possibility of *Realschule* graduates to transfer into gymnasial preparatory classes for the *Abitur*. By the end of the century they pointed to a surprising turnaround. The *Hauptschule*, once the school with the largest share of students, now attracted the fewest number, and the *Gymnasium*, having vaulted over the *Realschule*, now had taken the *Hauptschule*'s place. Statistics showed that of the total student population in the eighth grade, the percentage attending the *Gymnasia* had risen from 20.5 percent in 1960 to 29.3 percent in 1999. In the *Realschulen* that percentage had increased from 15.6 to 26.4, whereas in the *Hauptschulen* it had decreased from 63.6 to 22.6 percent. The *Gymnasium*, once an elite school for highly selected students, now attracted students from across the social spectrum.[59]

But while these changes have lessened the system's rigidity, they appear to have succeeded neither in improving the academic standing of *Hauptschule* students nor in corroborating the until then generally favorable international reputation of German public education. They also have not overcome what critics have always decried as the system's tendency to sort its students not only by presumed academic ability and performance, but also by social class, family income, and, in today's Germany with its large foreign-born populations, by nationality as well (table 5.1).[60] In addition, there are now indications that the social and ethnic divisiveness inherent in the three-pronged school structure effects the overall productivity of the German school system as measured in the outcomes of the OECD PISA tests of the year 2000.

These tests, carried out in thirty-two countries by the OECD Programs for International Student Assessment(PISA), came as an

Table 5.1 Socioeconomic Composition of Students in German Schools in 2000

School Type	Highest International Socioeconomic Index	Share of Working-class Families (%)	Share of Migration Families (%)	Mid-Level Reading Competence
Hauptschule	41.4	62.9	40.0	397
Realschule	48.3	42.5	20.3	500
Gymnasium	57.9	21.7	13.7	581

Source: Based on Table 9.12 in Jürgen Baumert et al., eds., *PISA 2000: Basiskompetenzen von Schülerinnen und Schülern im internationalen Vergleich* (Opladen: Leske und Budrich, 2001), p. 462.

unexpected shock to the German public and opened a new debate over the health of the German educational system.[61] The results showed that fifteen-year-old students in German schools ranked significantly below the OECD average competencies for reading, mathematics, and natural science as well as below the median values attained by students in schools in the United States. Among the thirty-two OECD countries Germany ranked twenty-first in reading competence and twentieth in mathematics and in the natural sciences. In the same categories the United States scored fifteenth, nineteenth, and fourteenth.[62] What particularly upset the German public, politicians, and educators was the fact that through its neglect of their weak students in the *Hauptschulen* the German segregated school system fared worse in its overall achievement ratings than the comprehensive schools of the United States. In the reading tests 10 percent of students in German schools fell below the lowest competency level. This compared to 6 percent for both the American and the OECD average. Similarly disturbing was the fact that in these tests the point distance between the 5 percent of the lowest scoring and the 5 percent of the highest scoring German students was the largest of all the participating countries. The German spread was 366, compared to the 328 for the OECD average and 349 for the United States. The PISA report states that this high spread was due above all, though not entirely, to weaknesses in the bottom sector.[63]

The results also illuminated the interrelationships that existed among the students' academic achievement, their social class, and the three-pronged school system. The PISA report states that

> it cannot be overlooked that 50 percent of the 15-year old students in the *Gymnasia* come from families of the upper civil service class and that, as one moves downward the social class scale, that percentage decreases to 10 for families of un- and semi-skilled workers. Correspondingly, in the *Hauptschule* the percentage increases from roughly 10 for children from the upper civil service class families to about 40 of those who come from families of unskilled workers.[64]

In the year 2000 the social class status of students attending German schools still corresponded roughly to the type of schools they enrolled in.

The same can be said for the ethnic or national backgrounds of students. Nearly 50 percent of the children from families in which both parents were born outside of Germany attended the *Hauptschule* or a vocational school, while only 15 percent of this group studied in a *Gymnasium*. Only 25 percent of the children from families in which both parents were born in Germany attended the *Hauptschule* or vocational school but 32 percent the *Gymnasium*. Those percentages did not differ much for children of families in which only one parent was born abroad.[65]

What German educational authorities have yet to fully acknowledge is that their schools, not unlike schools in the United States, serve many immigrant children whose parents were born abroad and who speak a language other than German in their homes (table 5.2). The linguistic competence of these immigrant children may often be below that which school authorities consider adequate for each of the different school types. The PISA report showed that only 30 percent of the students whose parents both were born abroad and in whose home a language other than German was most likely spoken reached the middle level of linguistic competency (Step 3), while of the students in families in which one or both parents were born in Germany, 60 percent did so.[66] Since it is known that a child's linguistic competency increases with time and experience, it is likely that *Schulwahl* at an age later than ten will increase a child's readiness for transfer into a *Realschule* or *Gymnasium*.

Table 5.2 Comparison of Immigrant Families, Germany and United States in 2000

Country	One Parent Born Abroad (%)	Both Parents Born Abroad	Families Speaking a Language Other than German at Home (%)
Germany	21.8	15.3	7.9
United States	19.5	13.6	10.8

Source: Based on Table 8.8 in Jürgen Baumert et al., eds., *PISA 2000: Basiskompetenzen von Schülerinnen und Schülern im internationalen Vergleich* (Opladen: Leske und Budrich, 2001), p. 348.

The evaluators of the German PISA report argued that because the lack of competence in the German language segregated and kept children of immigrant families in the *Hauptschule* and brought down the overall reading scores of students in German schools, an early identification of and support for these weak readers could materially improve the German schools' overall performance without lowering the prized superior reputation and achievement level of the *Gymnasia*.[67] If Germany would divert more resources to the *Hauptschule* and its students, it would also become more likely that more parents would out of their own choice opt for their children to attend a *Hauptschule* that will meet their children's needs and interests. They would not then have to accept a selection made for them by school authorities. And, if *Schulwahl* were to be postponed to a later year in a child's age, the performance level of immigrant children and their schools would be improved. Such changes, if carried out faithfully across the country, would do much to restore the vitality and effectiveness of German public schools.

Schulwahl and the related question of school choice, however, are not the only topics that come into play when the issue of parental participation in public school affairs is raised. In Germany's now ethnically quite diverse population, concerns over different social, religious, and intellectual values underlie and often prompt conflicts when parents insist on separating value-laden educational concerns from instructional matters. To be sure, the Federal Constitutional Court obliges state authorities to be mindful of the different values held by different groups in society and to refrain from attempting to indoctrinate students in specific patterns of social behavior. Based on the parental rights clause of the Basic Law (Article 6, Paragraph 2) the *Länder* have set up parent councils for grades and schools which send their delegates to similar councils on local, district, regional, and national levels.[68] But when taxpayers claim a right to participate in or object to curricular and educational decisions made by school authorities, they are reminded that they may not challenge the historically sanctioned, extensive state supervision over and direction of schools and education.[69] As Frank-Rüdiger Jach has pointed out, state supervision over curriculum and instruction, over character and personal development of students, and over the integration of students into society has been upheld in court decisions and in the legal as well as educational literature.[70] The parental voice in *Schulwahl* and other school affairs remains muted in most German schools.

Chapter 6

Retrospect and Outlook

Choice versus Compulsion in the National State

School choice, as we have seen, did not arise as an issue until the appearance of the modern national state. Then the question whether or not to send a child to public school became both a question for parents in their individual households and a debated public issue. It confronted everyone, parent and childless taxpayer alike, with the question to what extent a secular community—be it a town, a state, or a nation—could compel its citizens to hand over to them the education of their children. Could the state as *parens patriae* override the rights of parents as the determinators of their children's education and educate children in loco parentis? Was this a legal question, one to be decided in the last analysis by the crown as sovereign or by the people or their representatives in legislative assemblies and judicial bodies, or was it a moral issue to be settled by appeal to Scripture or natural law? Was it, finally, a matter to be turned over for decision to pedagogical expertise and professional educational wisdom? What kind of matter was this public education? One to be left to the discretion of individual parents or one to be decided by the community as a whole? Whose interest was it to serve, the individual's, the local community's, or society's in general? And if, as the name suggests, it was a matter for public disposition, how far could the public interfere in what many considered to be a most intimate sphere of private family life?

Once these questions had been raised and the debate had begun, the lines hardened between parents and citizens who insisted on their right to decide over their children's schooling and representatives of and spokespersons for the established school systems. Parents who had protested monopolistic practices of school authorities when they were still in the hands of landholders and churches, now objected to the edicts of governmental bodies. But hardly anywhere had school providers acknowledged such parental demands fully or even modestly. In premodern times landlords as employers or churches as spiritual supervisors had ignored, denied, or circumscribed parental protests just as governments did and do in modern states. The disputes over whose right and whose duty it is to assure the education of society's children is as old as organized society itself.

Today in Western societies it is considered to be a responsibility of the modern state to see to it that children are being educated, preferably in public schools, but, if parents insist and are willing to bear the costs, in private schools or in their own homes. To assure uniformity of outcomes, governments set standards concerning teacher qualifications, curricula, and textbooks. Parents who would like to be entirely free of governmental interference with their decisions concerning the place, time, manner, and content of their children's schooling will not be able to live in modern organized societies. In the past they have often sought such freedom in utopian communities, but even there they have encountered the ill will of their neighbors and have been forced to conform to the expectations and standards set by them.[1] No matter where we look, organized community life has imposed constraints on its members that make a totally free school choice unworkable and unrealistic.

The Historical Setting

Modern school choice as we know it today in the United States and in Germany had its beginning in the 1950s as part of wider demographic and cultural transformations that coursed through Western societies. School systems could not but mirror them. These transformations first burst into our consciousness with the worldwide student- and countercultural revolution of the 1960s. The assault on traditional authority in home, school, and public life led to a radical questioning of the entire notion of schooling and education.[2] The student rebels saw life as the opposite of organized instruction and enforced learning. They believed that their teachers and professors had nothing but antiquated wares to offer. Their elders, made uncertain by the righteousness of their accusers, in turn began to doubt their own role as teachers of the young and inexperienced.

But this was not all. A second assault followed the countercultural revolution. It was directed at the inherited and long since taken-for-granted belief in both the United States and Europe that the welfare state with its concern for society's poor and disenfranchised was here to stay. What came to be known in the United States as the Reagan revolution and in England as the Thatcher years appeared as pervasive distrust toward government as caretaker and provider. Government should make room for individual initiative and enterprise. It should abstain from championing the welfare of society's needy and leave such concerns to the market and private charity. This reaction to the welfare state spread among all the industrialized nations.

The combined effects of the two revolutions characterized our turn-of-the-century era. As Mark Lilla stated it well, "the cultural and Reagan revolutions took place within a single generation, and have proved to be complementary, not contradictory, events."[3] In culture and lifestyles as in politics and economics the two revolutions questioned and in many ways broke with established practices. The cultural revolutionists rejected what they considered calcified canons of outmoded respectability that stifled fresh growth with enforced obedience to lifeless convention. They wanted freedom for every lifestyle and Dionysian abandon and enjoyment of life. Their political brethren viewed the apparatus of government as equally stifling. Oppressive bureaucracy was the enemy. The demand was for a new release of energy. The common denominator was an incitement to individualistic, me-first thinking, resting on the assumption that, if only all of us take care of our own affairs, all else will be well. For better or worse, the appeal to self-interest, whether expressed in cultural acquirements or in material riches, has ever since pervaded the lifestyles of the industrialized nations.

Whether or not the proponents of school choice were aware of this, endorsed or rejected it, these new individualistic currents lent support to their cause. They fed the alarm, frustration, and anger

American parents, teachers, and students harbored over the educational deficiencies, racial, ethnic, and social inequalities, administrative inflexibilities, and financial insufficiencies of many public schools, particularly those in inner cities. That anger was by no means unjustified. Provoked as it was by racial, social, and gender discrimination and by neglect and deterioration of school buildings, it motivated parents and teachers who protested what Raymond Callahan had called "the cult of efficiency" and David Tyack had referred to as the drive toward "the one best system."[4] In both the United States and Germany parents and teachers detested a business mentality in education that sought financial bottom line efficiency and quantitative measurements of intelligence and "educational products" through standardized curricula and testing. They objected to school administrations that remained oblivious to discrimination against students, and encouraged invidious selectivity among them. They also believed that encouragement of idiosyncracies and individualistic interests rather than enforcement of conformity to lockstep movement captured the essence of education. They sought to achieve these desirable qualities within and separate and away from the public school system's intrusive and domineering supervision.

Opponents of school choice argued that the remedy to justified complaints against the public schools was not to leave them but to work to improve them. They pointed out that when anger and frustration led parents to withdraw their support from the public schools, it all too often followed that, however their children may have profited in alternative learning centers, they achieved their gains at the expense of the children of the wider community. In the United States opponents argued that school choice ran against public school ideology. The spirit of the public schools, they said, had as its object the welfare of that wider community and saw the common education of all of its children as an integral part of that welfare. This wider community should be seen as an indivisible whole. "No man is an island," they said, was the credo of the public schools. By forsaking the common good of this wider community, school choice not only endangered national unity and the common welfare, but also jeopardized the gains of racial integration won during the Civil Rights revolution.[5]

But whether we think of choice as alternatives in public schools or of choice as opting out of public schooling and choosing private education, the central issue has been, apart from the traditional reasons of ideology or religion, parental dissatisfaction with the existing state of public schooling. For minorities or majorities in the body politics, public school systems that remained unresponsive to citizen complaints furnished the major reason for school choice. When school administrations or political systems turned deaf ears to the opinions or wishes of parents, teachers, or students, then the public schools in fact ceased to be "the public's" schools and became the power centers of professional educators, administrators, or politicians.

Many parents felt that even the negative effects of school choice were preferable to a continuing adherence to the public schools. Love and concern for their children were powerful motivating forces that entered the picture. They obviously represented a constructive, positive force and supplied a vigorous vitality to the school choice movement. They counseled that one did not have to juxtapose the welfare of the whole with one of its parts, but that, in fact, the whole could benefit from the improvement of a part. There did not have to be an inevitable conflict between parents as private consumers of educational offerings and parents as public citizens. Parents have been both, and love of self did not necessarily conflict with love of others. As German *Reformpädagogik* and American progressive education in the early twentieth century had shown, private and public schooling could complement each other without detriment to either one. As the American charter school movement and German *Schulversuche* have demonstrated, there was no preordained conflict between school choice and the public schools.

The Swiss historian Jürgen Oelkers suggested that to put the current debate over school choice in its historical frame means to see it as but the latest phase of an ongoing dialectic between the defenders of national public school systems and their political and educational critics. As Oelkers put it, from the very beginning during the age of the Reformation the proponents of state systems of education saw them as means to serve state or national interests as these were defined by the representatives of society's dominant classes. They were to buttress mercantilist policies by providing an educated labor force and by training professionals in the churches, in law, administration, medicine, business, manufacturing, engineering, and the military. Necessarily, the advocates of such systems placed the viability of the school as institution and the stability of the school system as a whole at the center of their concerns. Their interests were primarily institutional and political. They wanted schools to create a unifying national language and culture and to instill loyalty to national ideals. Such aims, to be sure, also aided personal and local interests. But these benefits were incidental and, as commentators never failed to point out, in case of conflict, the personal and the local were to yield to the national interest.[6]

The political opponents of state school systems have always thought of schooling as an individual or local matter. They concerned themselves generally with community issues. Early on they recruited their supporters from among landowning patrons of rural schools, from clergymen committed to the creedal demands of their various faiths, from businessmen concerned primarily with the economic welfare of their communities, from intellectuals who saw the state as inherently antagonistic to the diverse interests and potential of individuals, and from among parents who feared state influence over the lives of their children. They argued that national and state school systems had been created artificially by statesmen and administrators; they had not developed naturally; they had not grown from below because of popular demand. These opponents became perennial advocates of educational reforms that were directed against the creation, development, and maintenance of state systems of education.

The educational critics, for their part, focused their concern on pedagogy in the classroom and on the best institutional and administrative conditions under which it could flourish. They spoke up for the interests of regimented youth and ignored parents. Philosophers and statesmen joined them as well as parents and citizens who looked upon schooling as a means to personal and community advance and welfare. They felt that the public school educators' emphasis on the national interest tended to overwhelm the legitimate claims of parents and pedagogues for the individual personhood of children and, inevitably, made schools "notoriously imperfect." "Nineteenth century educators," wrote Oelkers, "rarely rejected the school as social institution altogether, but they also did not accept it unconditionally. Their middle-way was the way of permanent reform, always disputed and notoriously uncertain."[7] At the end of the nineteenth and during the twentieth century, *Reformpädagogik* in Germany and progressive education in the United States had taken up the reform cause and made their contribution by placing the individuality of the child at the center of educational theory. Today, school choice is but another facet of the reform tradition.

As the historical chapters of this book have shown, religion had always played an important part in debates over school choice. In our Judaic-Christian civilization of the West, schooling has since the Middle Ages always carried a strong religious impulse. This became especially pronounced after the Protestant Reformation when Europe's secular leaders, seeking to remove the causes of religious strife and warfare, adopted the principle of *cuius regio, eius religio*, the rule that the confession of the sovereign would determine the confession of his subjects. Through the establishment of state churches sovereigns sought to compel confessional conformity within their territories and ultimately rely on religion as a means of gaining and confirming the loyalty of their subjects. Under these circumstances confessional religious instruction became a key subject in Prussia's public schools during the nineteenth century. It was mandated by the state school administration and avidly supported by the Evangelical Lutheran Church, the established church of Prussia. Prussia's other churches also benefitted from the arrangement as the country's public elementary schools were classified as either Catholic, Protestant, nondenominational Christian (the *Simultanschulen*), or Jewish, each obliged to offer religious instruction as part of their regular curriculum. By relying on religion as a criterion for classifying public schools, German school authorities provided opportunities for parents to exercise school choice on a religious basis. This possibility still exists in Germany today. In addition, the country's basic law gives parents the right to exempt their children from religious instruction or to send them to confession-free public schools in which religion is not part of the instructional program.[8]

In the United States, by contrast, the fear of the divisive effects of religious controversies and the antipathy to overbearing and all-pervasive governmental power, the latter reaffirmed and strengthened during the Revolution, persuaded the founders of the nation to keep religion outside the realm of government and to consider churches and synagogues private, not public, institutions. They erected what has come to be called "the wall of separation between church and state." In the First Amendment to the Constitution they prohibited established state churches and religious direction of public school activities. The last state to do away with its established churches was Massachusetts, which acted in 1833. Ever since, religious instruction cannot legally be offered in public schools and is considered a task of private institutions, such as churches, synagogues, parochial, and private schools.

Today's school choice advocates differ in their views on the role of religious instruction in public schools. Some accept the traditional way, barring such instruction. When in 1994 Peter Cookson asked for "a new educational covenant . . . that includes . . . managed public school choice," he left out private schools, and thus made no room for religious instruction.[9] Similarly, John Chubb and Terry Moe wrote in 1990 of a *public* school system that did not include private schools either.[10] Cookson, Chubb, and Moe restricted school choice to public schools because they believed the introduction of religion into schooling to be divisive and to lead to educational anarchy. It would, they felt, jeopardize the bonds of civic and national community. Their proposals thus did not meet the expectations of parents who desired religious instruction for their children.

But there were others, like John E. Coons and Stephen D. Sugarman in their 1978 book, who were willing to consider the inclusion in choice programs of private schools with tuition funds provided by public vouchers.[11] Again others, like Milton and Rose Friedman in their free market voucher approach, also demanded the inclusion of private schools, many of which would offer religious instruction. They argued that the expected benefits of an unrestricted free market voucher system outnumbered the threats it posed to the survival of a sense of civic and national community.[12]

Besides religion, ideological conviction, academic fashion, or theoretical considerations have been the motivating forces for school choice advocates. Thus school choice has become a vested concern of the Republican Party, of rational-choice philosophers, and of free enterprise economists. These proponents agreed with only minor differences that government influence over education is inherently wasteful and inefficient; that individual choice in educational matters will inevitably and automatically contribute to the general good; and that, compared to modern versions of town-meeting democracy, the marketplace is far superior in yielding optimal educational results.[13]

The historical development of private schooling, however, has not borne out the claims of the free market ideology of school choice. When the free market proponents celebrated economic motivations over political aspirations they seemed to assume that economic considerations had fueled private schooling and that monopolistic tendencies existed only within government bureaucracies and were unknown in competitive markets. Many private school educators, especially those in religious schools, will be astonished to hear that they have been or should have been inspired by economic rather than religious or educational concerns, particularly when they consider the low salaries generally paid to private school teachers and the tuition demanded of private school parents. Parents of children in religious schools will be surprised to learn that their schools are entirely free of any monopolistic taint. These ideological and theoretical claims of politicians and academics for school choice rest on shaky grounds when they elevate single-factor explanations—rational decision making of individuals aware of all relevant factors and assumed superiority of markets over democratic politics—to the ultimate criteria by which parents make educational decisions for their children. Parents are motivated in more complex ways.

This complexity of motivations of school choice advocates is not just a present-day phenomenon but has deep roots in the past. School choice proponents have come from various philosophical directions and, as a body, have represented diverse traditions of pedagogy. While the defenders of public schools in the main have subscribed to a utilitarian perspective that attempts to balance the concern for "the greatest good for the greatest number" with individual happiness, many school choice advocates have always considered a libertarian concern for individual autonomy a preeminent goal. When joined to a concern for equality of opportunity, the libertarian variety of school choice has placed and places great emphasis on merit, that is, academic ability. School choice advocates of social equality have stressed their concern for equity understood as equal outcomes and today favor such strategies as compensatory education.

A quite different tradition of school choice advocates has sought to balance the individual liberty aspects of school choice with communitarian concerns. This tradition has sought and seeks collective or social goals within the confines of relatively small communities, be they classrooms, neighborhoods, or towns, as contrasted with the larger entities of state or nation. This perspective has been dominant in the United States. It has its roots in colonial days and in nineteenth-century rural America where people in their towns and villages insisted on local control. They did this because they were jealous of their rights to republican self-government and determined to preserve for themselves the right to resist outside taxation. Joining local communities in their resistance were religious and racial groups. The establishment in the nineteenth century of a separate parochial school system of American Catholics has been a case in point, as have been the efforts of African-American parents in the twentieth century to introduce community control, magnet and charter schools as well as voucher plans in the country's inner cities.

In Germany school choice proponents have relied heavily on the humanistic tradition of concern for the individual child and its personal development in the context of the child's or student's local community. This tradition dates back to the writings of such late-eighteenth- and early-nineteenth-century philosophers and educators as Wilhelm von Humboldt, Johann Friedrich Herbart, and Friedrich Schleiermacher. It has sought to provide educational settings in which self-actualization and education as therapy—the child-centered school of the Progressives—can become a reality.[14]

In all these cases we have seen school choice advocates as spokespeople of what today we call civil society. As such they have found their antagonists in the representatives of the state organized in state educational administrations or state and national governments. There have been, however, significant differences in the way this confrontation between civil society and government has played out in the United States and Germany. In the United States, the dominance throughout the nineteenth century of local control of public education has established the communitarian perspective as a viable tradition not easily to be ignored or discarded. This perspective lies at the heart of the American concept of civil society. It has enabled civil society to maintain itself as a partner equal in staying power with state and federal government. Whether today under the growing dominance of the federal government over public education it will be able effectively to maintain its position and oppose governmental regulations and supervision remains to be seen. While school choice advocates in the United States today may relish the many governmental pronouncements supporting their cause, the imposition of standard-based reforms and testing programs under the "No Child Left Behind Act" may well give them pause.

In Prussia, and later in Germany, civil society never succeeded in appearing as an equal partner with government. The advances it secured—the establishment of private schools and the school reforms of the 1960s and 1970s—remained under the authority of the state and were placed under government regulation and supervision. Private schools were never recognized as market institutions. As they have been subjected to state supervision and frequently received state financial aid they were and are viewed as substitutes for public schools and differ little from them. They are sometimes referred to as belonging to a "third," nonprofit or voluntary, sector which itself is often seen as an integral part of the state. The reforms of the 1960s and 1970s were concerned less with private than with public schools. They revolved around parental involvement in public school government and increase in the autonomy of public schools. As all of these undertakings fall under the supervising authority of the state as defined in Article 7, Paragraph 1 of the German Basic Law of 1949, German civil society has remained a very junior and legally unrecognized partner in its relationship to the state.

Choosing among Public and Private Schools

Once public schooling had become the norm in the United States and in Prussia, school choice involved above all a parental decision whether a child was to attend a private or a public school. In the United States, popular opposition made ineffective essays and proposals in favor of a state system of common education that had been submitted by philosophers and pedagogues after the Revolution. The people in their localities insisted upon providing for and governing their own common schools. Even when by the middle of the nineteenth century state systems of public education were established, popular sentiment in favor of local control over public schools persisted. Private schools, too, continued to flourish, especially those on the secondary level.[15] When toward the end of the nineteenth century public, state supervised education had become the rule and included ever larger number of secondary schools, parents still could and did avail themselves of private schools for their children.

In Prussia philosophers, statesmen, and educators debated at the turn from the eighteenth to the nineteenth century whether common schooling for their country's children would be most beneficially performed under state direction or, when in private hands, supported only by local personal and civic initiatives. State legislation and administrative regulations, in some cases proposed and executed by the same men who had participated in the debate, settled the issue in favor of public control of common schooling. Private schools were thenceforth assigned a distinctly secondary role. Still, private schools continued and continue to exist in Germany to the present day and have permitted parents to choose between them and the public system.

In the United States the Constitution had not placed education under federal supervision. To the contrary, it had omitted education among the enumerated powers of the Congress and, in the Tenth Amendment, had assigned it to the states or to the people.[16] So it was the people who during the new nation's first decades created their own common schools in their localities and believed that these were indeed their schools which were to be run by them as they saw fit. The closeness of locally elected school officials to parents mattered more to them than the arguments heard in philosophical contests over the education appropriate for citizens of a republic. This was to remain true also when during the 1830s and later decades various states established departments of public instruction. Still, citizens and parents preferred their own elected local to state school administrators. They wanted to have a say in the supervision of their local public schools. That concerned them more than a debate over public versus private school ownership.

In Prussia and later Germany things were not that different. Relatively few parents exercised their right to choose between a public or private school for their child. The state's supervisory authority, originally established in the General Land Law of 1794 and extended over both public and private schools, did not encourage that choice. In the nineteenth century, state officials as well as representatives of civil society who as municipal officeholders and members of school societies were responsible for the financing of public schooling, accepted the state's authority as given. Because Prussia's churches were considered public institutions, their pastors and priests also spoke as public officials. Thus the issue for most parents was not how to chose between a public or a private school, but how to find in their locality a public school most appropriate for their child's future. To that concern local public school officials were more likely to respond than officials in the state administration.

In both countries church-sponsored private schools have come to claim the largest number of private schools and students. American students enrolled in religious schools account for 84 percent of all private school students. In Germany that figure comes to about 75 percent. In both countries schools under the sponsorship of the Catholic Church rank first in enrollments among private religious schools. Their students amount to about 50 percent of all children enrolled in private schools. Religious faith and the conviction that through religious instruction their children will gain lifelong guidance in ethics and morality as well as a reputed strengthened appreciation of academic learning prompt parents to send their children to schools under religious sponsorship.

The roles of church-sponsored private schools, however, have been quite different in the United States and in Prussia. In the United States the separation of church and state and the post–Civil War agitation for a national education system had provided a strong incentive for educators, churches, and synagogues to found nonpublic schools. The secular and especially Protestant nativist campaigns directed against the Catholic Church at the end of the nineteenth century backfired and encouraged not only the establishment of a Catholic education system but of various non-Catholic religious schools as well. At the end of the twentieth century these religious schools together with secular private schools have brought the share of school children enrolled in them to about 10 percent of the country's total elementary and secondary school population, not quite double the percentage it had reached in Germany.

In Prussia, and later in Germany, public schools have been for the most part confessional institutions offering religious instruction of the major Christian and, in the nineteenth century, Jewish congregations. There was, then, no urgency for the churches and synagogues to open parochial schools of their own. But Jewish public schools were closed during the Third Reich, and many Christian public schools gave up their confessional character during the 1960s. It was then that parents who desired religious instruction for their children as well as those who complained about the public schools' unresponsiveness to their concerns, began to look for private schools. Still, by the end of the twentieth century, the share of German school children enrolled in both religious and secular private schools did not exceed 6 percent and the share of German private schools among all general education schools was only 5 percent. For Europe—where in Great Britain the percentage of private schools stood at 25, in France where it reached 30, and in the Netherlands where it ran as high as 70—this percentage was remarkably low.[17]

The appeal of secular private schools rests primarily on their reputation for personalized attention to their students' individual needs and interests. In the United States, some private schools under secular sponsorship follow the pedagogical traditions of European Montessori and Waldorf schools and others are native products of the progressive movement. They attract parents who seek a more individualized education for their children than most public schools can provide. In Germany secular private schools are known for their various pedagogical specializations that range from the country boarding schools of international academic and outdoor education fame to schools for students with learning disabilities.

Present-Day School Choice in the United States

In the United States changes in public school policies and school reform have been the subject of nearly constant initiatives and debates. They included parent- and student-initiated alternative programs in public schools but went beyond these to many of the innovations I discussed in chapter 5. They were accompanied by federal legislation for increased federal funding of public education and, as an inevitable consequence, for strengthened federal controls as well. The future of school choice issues in the United States, issues that had their origin in the American commitment to local control in education, is now inextricably bound up with the role of the federal government in education.

At the beginning of the twenty-first century school choice supporters in the United States find themselves in a strange position, confronted with contradictory signals from state and federal governments. On the one hand, there is considerable support from public authorities for the idea of choice and alternatives in education. Conservative legislators praise the value of free market competition in schooling and decry what they consider the inefficiencies of democratically elected school governments and public administrations. On the other hand, their penchant for performance-based accountability has led state and federal legislators to impose on the country's schools high-stake tests that function like a straight-jacket on curricula and daily classroom teaching. Apparently the proponents of these views are quite oblivious to the contradictions inherent in these policies.

Given these contradictions, parents will have to ask whether the presence of the federal government in public education will bring them a much welcomed assistance in their struggles with unresponsive local school administrations or whether the demands for accountability through the imposition of standards and tests will strike them as an unwarranted and unwanted forced intrusion into their children's classrooms. School choice in its original meaning had given them opportunities to shape their children's academic careers and a voice in the operation of their children's schools. It had encouraged progressive education reforms and allowed wide room for experimentation and fresh approaches in the classroom. Standard-based instruction and mandatory testing, on the other hand, vitiate these reforms. They inhibit a teachers' spontaneity of teaching and destroy the students' love of learning. Inevitably, they shortchanged creative subjects of the curriculum and lead to teaching to the test.

How standard-based instruction and mandatory high-stakes testing affect teaching and learning is vividly described by Dale and Bonnie Johnson, two college education professors who spent a sabbatical year teaching in a poverty-stricken rural Louisiana elementary school:

> The use of high-stakes tests is changing what goes on in classrooms to the detriment of the arts, problem solving, creativity, and the joy associated with learning and discovering. . . . Time is spent cramming the materials covered in the tests to the neglect of entire subjects such as science, social studies, art, drama, and music.

Their concluding recommendation urges that "state legislatures, boards of education, and school district authorities must back off from their commitment to raise tests scores at the expense of providing children a well-rounded, well-grounded education across the curriculum."[18] What started out as education reform now tends to lead to the reimposition of drill and regurgitation of memorized information.

This is not to say that the demand for accountability from public schools is unjustified. It is not. But to have it come to mean testing alone is to pervert it. In its early stages, wrote Richard Elmore, accountability included, "in addition to tests, portfolios of students' work, teachers' evaluations of their students, student-initiated projects, and formal exhibitions of students' work." It was then part of a wider school reform movement that was entirely compatible with school choice reforms. But in its mandatory single-test version accountability, states Elmore, "is mutating into a caricature of itself."[19]

The imposition on the nation's schools of single-test based accountability systems had come through the "No Child Left Behind Act of 2001." In this act the federal government had announced its support for school choice and offered funds to aid parents in their search for public charter schools for their children. Much to many parents' dismay, and despite its often expressed support for a greatly expanded school choice program, the Bush administration has not included private schools in its purview. The reason, we may assume, was its unwillingness to antagonize the owners of private and religious schools who feared that if they were to receive federal funds, they would have to accept federal accounting requirements and thus lose their schools' independence from governmental controls. The public school establishment's resistance to outright government support of private schools was another factor as well.

The act also did not offer any support to parents who had hoped for federal endorsement of voucher programs that would allow them to be supported with federal funds for the attendance of their children at private schools. President Bush, knowing that many state constitutions forbade the expenditure of public funds in support of religious institutions and activities, stated that it was not the federal government's prerogative to "impose a school voucher plan on states and local jurisdictions."[20] He was also well aware that the use of vouchers was a much disputed and debated issue across the country which still awaited a Supreme Court ruling on its constitutionality.

That ruling came in the summer of 2002 in a five to four decision in the *Zelman v. Simmons-Harris* case. It declared constitutional the use of vouchers to provide "educational assistance to poor children in a demonstrably failing public school system." The majority opinion held that a program of government funds provided through vouchers could not be challenged under the establishment clause of the First Amendment as long as it was "neutral with respect to religion" and the funds were given to religious schools as a result of the recipients' "own genuine and independent private choice." Of the five justices joining in the majority decision, two filed concurring opinions. Of the four who made up the minority, three wrote their own dissenting opinions. In one of these, three of the minority justices joined; in a second, two colleagues did.[21]

The divisions among the justices of the Court and the number of concurring and dissenting opinions underscore the divisiveness of the issue. As Supreme Court decisions are not chiseled in stone to remain unvaryingly prescriptive throughout the ages and as they reflect and give legal expression to the ever-changing social and political currents pulsing through society, it is difficult to predict what long-range effect the *Zelman* decision will have on the future politics of vouchers. Nonetheless, the case is important enough to be treated here in some detail. It illuminates the sensitive and problematical nature of school choice as it pivots on the fundamental issue of the relationship between church and state which, in turn, hinges on the Court's reading of the First Amendment's establishment clause.

In his dissenting opinion Justice Souter reviewed the Court's changing interpretations of the establishment clause. What ties the establishment clause to parental school choice is the funneling of public funds to private, and specifically religious, institutions. Justice Souter referred his colleagues to the *Everson* case of 1947 which revolved around that issue. At that time Justice Black, speaking for the majority, had declared constitutional a New Jersey statute that authorized school districts to use tax money to compensate parents for bus transportation of their children to school, even when some of the children were bussed to parochial schools. The state, said Justice Black, did not contribute money to schools. It did no more "than to provide a general program to help parents get their children, regardless of their religion, safely and expeditiously to and from accredited schools." Then, to make sure that his ruling did not in any way question or overturn the traditional interpretation of the establishment clause, he added: "The First Amendment has erected a wall between church and state. That wall must be kept high and impregnable. We could not approve the slightest breach. New Jersey has not breached it here."[22] These are the words Justice Souter pointed to when he wrote in his 2002 dissent: "Although the [1947] Court split, no Justice disagreed with the basic doctrinal principle that 'no tax in any amount . . . can be levied to support any religious activities or institutions . . .' "

Justice Souter's historical account then made clear that by 1983 the Court had come to distance itself from that "no tax money for religious institutions" principle. Though the Court's focus had remained "on what the public money bought when it reached the end point of its disbursement," it had become blurred in a series of intervening cases.[23] The Court had found it difficult to draw the line that defined the divertibility of tax funds and to determine whether public funds, when given to religious schools, could or would be used for religious purposes. By 1983, in *Mueller v. Allen*, the Court then adopted two new criteria: neutrality of fund availability and private choice. Justice Rehnquist explained the meaning of neutrality. Referring to a tax deduction granted by a Minnesota statute he said: "Most importantly, the deduction is available for educational expenses incurred by all parents, including those whose children attend public schools and those whose children attend nonsectarian private schools or sectarian private schools." As to private choice, he said, ". . . under Minnesota's arrangement public funds become available only as a result of numerous private choices of individual parents of school-age children."[24] These two conditions had now become the Court's new standard. As Chief Justice Rehnquist reiterated in the 2002 voucher case his opinion given in 1983 in *Mueller v. Allen*, neutrality of fund availability and private choice were sufficient to overcome the hurdle of the establishment clause. The fact that substantial public funds flowed into religious coffers no longer seemed to matter.

As the Court stressed that the tuition aid provided through vouchers favored low income and minority families whose children went to inner-city public schools in the Cleveland City School District and who now had a choice of other schools, the direction of its argument became apparent. The Cleveland district had, for more than a generation, "been among the worst performing public schools in the nation." Parents now had a choice. Those who wanted to keep their children in their regular public schools were to receive tutorial aid. Those who wanted to send their children to private religious and nonreligious schools and those who preferred to have their children taught in regular Cleveland public schools, in public schools in adjacent districts, or in charter and magnet schools either paid no tuition or would receive state aid. As all of these schools were available to Cleveland parents, Justice Rehnquist could say that the program permitted "individuals to exercise genuine choice among options public and private, secular and religious. The program is therefore a program of true private choice." Given the history of the Court's decisions outlined by Justice Souter, the majority's conclusion followed unavoidably: The Cleveland program was constitutional.

Justice Souter, however, did not agree. He found that the majority's use of the neutrality criterion was illogical in as much as the Cleveland voucher program counted all schools receiving public funds, including regular public schools, as participants. Under the voucher program, however, regular public schools were allowed to receive no more than $324 per child to support extra tutoring, whereas the mostly religious tuition voucher schools could receive up to $2,250 per child. There was, then, no neutrality of fund availability between regular public and tuition voucher schools.

Justice Souter found the majority's application of the free choice criterion equally unconvincing. As Souter saw it, free choice referred to a parent's ability to spend the scholarship money in either a secular or religious school of one's choosing, not in having available an array of schools open to anyone willing to attend a public school. The alternatives that choice required, he stated, was between secular and religious private schools. In the Cleveland case, of the 53 private schools accepting vouchers in 1999–2000, 46 were religious schools, attended by 96.6 percent of all voucher recipients. But this attendance ratio had little to do with a preference for religion. Parents listed educational opportunity and safety as their chief reasons. Besides, the $2,500 cap on voucher payments tended to curtail enrollments in secular private schools that charged as much as $4,000 tuition, compared to the average tuition of $1,592 at Catholic schools. As a result, stated

Souter, "for the overwhelming number of children in the voucher scheme, the only alternative to the public schools is religious."

Summing up his dissent, Justice Souter pointed out that "the scale of the aid to religious schools approved today is unprecedented" and "every objective underlying the prohibition of religious establishment is betrayed by this scheme." It showed disrespect for students' freedom of conscience and infringed on the churches' freedom to give preference to students and teachers of their own denomination. "When government aid goes up," wrote Souter, "so does reliance on it; the only thing likely to go down is independence." And finally Souter was joined by justices Stevens and Breyer in his concern for an increase in religious strife and social conflict. Government funding of private school teaching of religion, concluded, Justice Breyer,

> is far more contentious than providing funding for secular textbooks, computers, vocational training, or even funding for adults who wish to obtain a college education at a religious university. . . . [It] is far more divisive than state property tax exemptions for religious institutions or tax deductions for charitable contributions . . .

In the dissenting justices' opinion, the majority decision in the *Zelman* case betrayed the long established American tradition of separation of church and state. It constituted a threat to social peace and bequeathed a questionable legacy to the cause of school choice.

At the beginning of the twenty-first century, then, school choice in the United States is as controversial as ever. The viability of charter schools goes up and down with the fate of individual schools, district policies, and state legislation. The voucher issue, as we have just seen, gets a disputed boost by a Supreme Court decision, yet it has so far failed to gain the support of federal legislation. Its fate depends to a large degree on the performance of the nation's public schools. If school reforms succeed in bolstering the public's confidence in its public schools and overcome the misery and neglect of poverty-stricken rural and inner-city schools, the calls for school choice will subside. If federal programs continue to standardize education, erode the vitality and spontaneity of classroom teaching and learning, the demands for school choice will rise correspondingly. In the last analysis, the fate of American schools will depend on the wisdom of the legislators in state houses and the national Congress.

Notes

Introduction School Choice: A Brief Overview

Notes to Pages 1–3

1. See as examples for Germany, Frank-Rüdiger Jach, *Schulvielfalt als Verfassungsgebot* (Berlin: Duncker & Humblot, 1991); for the United Kingdom, G. Walford, ed., *School Choice and the Quasi-Market* (Wallingford: Triangle Books, 1996); for the United States, Peter W. Cookson, *School Choice: The Struggle for the Soul of American Education* (New Haven and London: Yale University Press, 1994); and for Australia and New Zealand, Geoff Whitty, Sally Power, and David Halpin, *Devolution and Choice in Education: The School, the State and the Market* (Buckingham, England: Open University Press, 1998). See also the comparative OECD Report with examples from Australia, England, New Zealand, the Netherlands, Sweden, and the United States, Centre for Educational Research and Innovation, *School: A Matter of Choice* (Paris: Organization for Economic Co-operation and Development, 1994).
2. John E. Chubb and Terry M. Moe, *Politics, Markets, and America's Schools* (Washington, DC: Brookings Institution, 1990), p. 219.
3. On this see Diane Ravitch, *The Troubled Crusade: American Education 1945–1980* (New York: Basic Books, 1983), pp. 308–311.
4. For the German side on this see Günter Püttner, "Die Sprengelpflicht," *Recht der Jugend und des Bildungswesens: Zeitschrift für Schule, Berufsbildung und Jugenderziehung* 40 (1992): 230–235. In the United States, the neighborhood school concept expresses a similar practice, albeit in less forceful and legalistic language.
5. Three noteworthy exceptions are William W. Cutler, III, *Parents and Schools: The 150–year Struggle for Control in American Education* (Chicago: University of Chicago Press, 2000); Otto F. Kraushaar, *Private Schools: From the Puritans to the Present* (Bloomington, IN: Phi Delta Kappa

Educational Foundation, 1976); and Nancy Beadie and Kim Tolley, eds., *Chartered Schools: Two Hundred Years of Independent Academies in the United States, 1727–1925* (New York: RoutledgeFalmer, 2002).

6. See Bernard Bailyn, *Education in the Forming of American Society* (New York: Vintage, 1960); and Lawrence Cremin, *The Wonderful World of Ellwood Patterson Cubberley* (New York: Teachers College Press, Columbia University, 1965).
7. One outstanding exception is Wayne E. Fuller, *The Old Country School* (Chicago: University of Chicago Press, 1982).
8. See Hermann Röhrs and Volker Lenhart, *Progressive Education Across the Continents* (Frankfurt am Main: Peter Lang, 1995).
9. See Jürgen Oelkers, *Reformpädagogik: Eine kritische Dogmengeschichte* (Weinheim and München: Juventa Verlag, 1989) and my review essay of Röhrs and Lenhart, *Progressive Education*, "Toward a Theory of Progressive Education," *History of Education Quarterly* 37 (Spring 1997): 45–59.
10. Arbeitsgruppe Bildungsbericht, Max Planck Institut für Bildungsgeschichte, *Das Bildungswesen in der Bundesrepublic Deutschland* (Hamburg: Rowohlt, 1990), pp. 140–143.
11. Frank-Rüdiger Jach, *Schulverfassung und Bürgergesellschaft in Europa* (Berlin: Duncker & Humblot, 1999), p. 253.
12. Hellmut Becker, "Die verwaltete Schule (1954)," in *Quantität und Qualität: Grundfragen der Bildungspolitik* (Freiburg im Breisgau: Verlag Rombach, 1962), p. 148.
13. See my "An International Brotherhood of Whigs: Nineteenth Century School Reformers in the United States and Prussia," *Education Research and Perspectives* 21, no. 2 (2004): 1–10.
14. Henry Geitz, Jürgen Heideking, and Jurgen Herbst, eds., *German Influences on Education in the United States to 1917* (Cambridge: Cambridge University Press, 1995).

1 Beginnings

1. See Jon Teaford, "The Transformation of Massachusetts Education, 1670–1780," reprinted in *The Social History of American Education*, ed. B. Edward McClellan and William J. Reese (Urbana: University of Illinois Press, 1988), p. 35.
2. James B. Conant, *Thomas Jefferson and the Development of American Public Education* (Berkeley: University of California Press, 1962), p. 110.
3. Pierre Samuel du Pont de Nemours, *National Education in the United States of America* (Newark: University of Delaware Press, 1923), pp. 147–148.

4. For more on these various proposals and for the locations of the quoted passages see Frederick Rudolph, *Essays on Education in the Early Republic* (Cambridge: Harvard University Press, 1965), pp. 10, 14, 17, 64, 66–68, 82, 113, 138, 210, 318–319.
5. See on this Siobhan Moroney, "Birth of a Canon: The Historiography of Early Republican Educational Thought," *History of Education Quarterly,* 39 (Winter 1999): 476–491.
6. In this and the preceding paragraph I rely heavily on Carl F. Kaestle and Maris A. Vinovskis, *Education and Social Change in Nineteenth-Century Massachusetts* (Cambridge: Cambridge University Press, 1980), pp. 9–27.
7. See my *The Once and Future School: Three Hundred and Fifty Years of American Secondary Education* (New York: Routledge, 1996), pp. 42–45.
8. See my *From Crisis to Crisis: American College Government 1636–1819* (Cambridge: Harvard University Press, 1982), pp. 139, 168–169.
9. Regent Minutes, *Journals of the Legislature of the State of New York, Assembly Journal,* February 28, 1832.
10. See Frank C. Abbott, *Government Policy and Higher Education: A Study of the Regents of the University of the State of New York, 1784–1949* (Ithaca, NY: Cornell University Press, 1958), *passim,* and my "The Regents of the University of the State of New York, 1784–1920: Secondary Education Emerges in the New Nation," in Antonio Novoa, Marc Depaepe, and Erwin V. Johanningmeier, eds., "The Colonial Experience in Education," *Paedagogica Historica,* Supplementary Series 1(1995): 317–333.
11. Committee on Colleges, Academies, and Common Schools, Regent Minutes, *Assembly Journal,* March 6, 1821.
12. See my *The Once and Future School,* pp. 55–56.
13. See my *The Once and Future School,* pp. 58–60.
14. See Lawrence A. Cremin, *American Education: The National Experience 1783–1876* (New York: Harper & Row, 1980), p. 151.
15. E. G. West, *Education and the State: A Study in Political Economy,* 3rd ed., rev. and exp. (Indianapolis: Liberty Fund, 1994), pp. 297–391.
16. David B. Tyack, *The One Best System: A History of American Urban Education* (Cambridge: Harvard University Press, 1974), pp. 33–34.
17. For an account and analysis of the 1840 school board election see Kaestle and Vinovskis, *Education and Social Change,* pp. 213–232.
18. Editorial in *The Massachusetts Teacher* 4(1851): 289–291.
19. Henry Barnard, *Report on the Condition and Improvement of the Public Schools of Rhode Island* (Providence, RI, B. Cranston and Co. 1845), p. 50, reprinted in *Henry Barnard on Education,* ed. John S. Brubacher (New York: McGraw-Hill, 1931), p. 276.

20. Kaestle and Vinovskis, *Education and Social Change*, pp. 212– 213, 208.
21. Kaestle and Vinovskis, *Education and Social Change*, pp. 225–235.
22. See Maris A. Vinovskis, *The Origins of Public High Schools: A Reexamination of the Beverly High School Controversy* (Madison: University of Wisconsin Press, 1985), p. 111.
23. See Carl F. Kaestle who in his *Pillars of the Republic: Common Schools and American Society, 1780–1860* (New York: Hill and Wang, 1983), p. 147, calls localism "one of the most enduring and pervasive sources of conflict in American educational history."
24. Carl F. Kaestle, *The Evolution of an Urban School System: New York City, 1750–1850* (Cambridge: Harvard University Press, 1973), pp. 68–71.
25. Diane Ravitch, *The Great School Wars: New York City, 1805–1973: A History of the Public Schools as Battlefield of Social Change* (New York: Basic Books, 1974), pp. 36, 59, 43, and *passim*.
26. Kaestle, *The Evolution of an Urban School System*, p. 149.
27. Vincent P. Lannie, *Public Money and Parochial Education: Bishop Hughes, Governor Seward, and the New York School Controversy* (Cleveland: Press of Case Western University, 1968), especially Chapter 12.
28. David Tyack and Elisabeth Hansot, *Managers of Virtue: Public School Leadership in America, 1820–1980* (New York: Basic Books, 1982), p. 78.
29. Leonhard Froese and W. Krawietz, eds., *Deutsche Schulgesetzgebung (1763–1952)*, 2nd ed. (Weinheim: Julius Beltz, 1968), pp. 91, 94, 105–111. See also Heinrich Lewin, *Geschichte der Entwicklung der preußischen Volksschule* (Leipzig: Dürr'sche Buchhandlung, 1910), pp. 42–91.
30. Froese and Krawietz, *Deutsche Schulgesetzgebung*, p. 23, and Lewin, *Entwicklung*, pp. 142–146.
31. See Franzjörg Baumgart, *Zwischen Reform und Reaktion: Preußische Schulpolitik 1806–1859* (Darmstadt: Wissenschaftliche Buchgesellschaft, 1990), p. 55.
32. See Hans Heckel, *Schulrecht und Schulpolitik: Der Einfluß des Rechts auf die Zielsetzung und den Erfolg in der Bildungspolitik* (Neuwied/Berlin: Hermann Luchterhand Verlag, 1967), p. 49.
33. Cited in Frank-Michael Kuhlemann, *Modernisierung und Disziplinierung: Sozialgeschichte des preußischen Volksschulwesens 1794–1872* (Göttingen: Vandenhoeck & Ruprecht, 1992), p. 64. In German the relevant terms are: While schools are *Veranstaltungen des Staates*, they are not *Staatsanstalten*.

34. Kuhlemann, *Modernisierung*, p. 86.
35. Baumgart, *Zwischen Reform und Reaktion*, pp. 66–78.
36. Manfred Heinemann, *Schule im Vorfeld der Verwaltung: Die Entwicklung der preußischen Unterrichtsverwaltung von 1771–1800* (Göttingen: Vandenhoeck und Ruprecht, 1974), p. 33.
37. Heinemann, *Schule im Vorfeld*, pp. 24–25.
38. Zedlitz's proposals, "Vorschläge zur Verbesserung des Schulwesens in den Königlichen Landen," were first published in the *Berlinische Monatsschrift* 10 (1787): 97–116. They have been conveniently reprinted in Christa Berg, ed., *Staat und Schule oder Staatsschule? Stellungnahmen von Pädagogen und Schulpolitikern zu einem unerledigten Problem, 1781–1889* (Königstein: Athenäum, 1980), pp. 1–9. They are discussed in Heinemann, *Schule im Vorfeld*, pp. 152–156.
39. See Baumgart, *Zwischen Reform und Reaktion*, p. 17.
40. Heinemann, *Schule im Vorfeld*, p. 168.
41. Baumgart, *Zwischen Reform und Reaktion*, pp. 22–24.
42. See Berg, *Staat und Schule oder Staatsschule?* pp. 10–21.
43. Berg, *Staat und Schule oder Staatsschule?*, pp. xviii–xix and 22–36.
44. "Ideen zu einem Versuch, die Grenzen der Wirksamkeit des Staats zu bestimmen (1792)," in Wilhelm von Humboldt, *Werke in Fünf Bänden*, ed. Andreas Flitner and Klaus Giel (Darmstadt: Wissenschaftliche Buchgesellschaft, 1960), 1: 56–233. (Section on public education in Chapter 6 appears on pp. 103–109.) See also the English language edition, J. W. Burrow, ed., *The Limits of State Action* (Indianapolis: Liberty Fund, 1993), pp. 46–52.
45. Quoted in Herbert Scurla, *Wilhelm von Humboldt: Werden und Wirken* (Berlin: Verlag der Nation, 1970), pp. 81 and 201.
46. Cf. the interpretation advanced in Ursula Krautkrämer, *Staat und Erziehung: Begründung öffentlicher Erziehung bei Humboldt, Kant, Fichte, Hegel und Schleiermacher* (München: Johannes Berchmans Verlag, 1979), p. 77.
47. Cited in Kuhlemann, *Modernisierung und Disziplinierung*, p. 76.
48. See "Königsberger und Litauischer Schulplan," in Humboldt, *Werke*, 4: 181–184, and "Bericht an Altenstein über die Finanzgrundsätze der Sektion," in Humboldt, *Werke*, pp. 283–284.
49. Cf. Ursula Krautkrämer, *Staat und Erziehung*, p. 77.
50. See Frederic Lilge, *The Abuse of Learning: The Failure of the German University* (New York: Farrar, Straus and Giroux, 1948), p. 52.
51. See Johann Gottlieb Fichte, "Reden an die deutsche Nation," in *Sämmtliche Werke*, ed. I. H. Fichte (Berlin: Veit, 1846), 7: 271–427.

52. Berg, *Staat und Schule oder Staatsschule?*, pp. xix–xi; Johann Friedrich Herbart, "Über Erziehung unter öffentlicher Mitwirkung, (1810)," in *Berg, Staat und Schule oder Staatsschule?*, pp. 37–45.
53. F. E. D. Schleiermacher, "Über den Beruf des Staates zur Erziehung (1814)," in *Ausgewählte pädagogische Schriften*, selected by Ernst Lichtenstein (Paderborn: Verlag Ferdinand Schöningh, 1959), pp. 28–29.
54. Kuhlemann, *Modernisierung*, p. 77.
55. Kuhlemann, *Modernisierung*, pp. 80–82, 348–349.

2 The Systematization of Public Education

1. See on this William J. Reese, *Power and the Promise of School Reform: Grassroots Movements during the Progressive Era* (Boston: Routledge & Kegan Paul, 1986), p. 4.
2. Quoted in Stanley K. Schultz, *The Culture Factory: Boston Public Schools, 1789–1860* (New York: Oxford University Press, 1973), pp. 269–270.
3. Louis R. Harlan, *Separate and Unequal: Public School Campaigns and Racism in the Southern Seaboard States 1901–1915* (New York: Atheneum, 1968), p. 3.
4. Jacqueline Jones, *Soldiers of Light and Love: Northern Teachers and Georgia Blacks, 1865–1873* (Athens: University of Georgia Press, 1992), p. 14.
5. Carter Godwin Woodson, *The Mis-Education of the Negro* (Washington, DC: Associated Publishers, 1933), p. 28, and his *The Education of the Negro prior to 1861*, 2nd ed. (Washington, DC: Associated Publishers, 1919), p. 17.
6. W. E. B. DuBois, *Black Reconstruction in America* (New York: Russell & Russell, 1935), pp. 638 and 645.
7. See Christopher M. Span, "Alternative Pedagogy: The Rise of the Private Black Academy in Early Postbellum Mississippi, 1862–1870," in *Chartered Schools: Two Hundred Years of Independent Academies in the United States, 1727–1925*, ed. Nancy Beadie and Kim Tolley (New York and London: RoutledgeFalmer, 2002), pp. 211–227.
8. From the *Congressional Globe*, 41st Cong., 1st sess., 1879, 49, pt. 1: 1735, cited in Lloyd P. Jorgenson, *The State and the Non-Public School, 1825–1925* (Columbia: University of Missouri Press, 1987), p. 137.
9. Henry Wilson, "New Departure of the Republican Party," *The Atlantic Monthly* 27 (January 1871): 119–120. See also Jorgenson, *The State and the Non-Public School, 1825–1925*, pp. 137–138.

10. Cited from *Catholic World* 13 (April 1871): 1–14.
11. Otto F. Kraushaar, *American Nonpublic Schools: Patterns of Diversity* (Baltimore: Johns Hopkins University Press, 1972), p. 14.
12. *Everson v. Board of Education*, 330 U.S. 1, 16 (1947). Speaking for the majority Justice Black declared constitutional a New Jersey statute that authorized school districts to use tax money to compensate parents for bus transportation of their children to school, even when some of the children were bussed to parochial schools. The state, said Justice Black, did not contribute money to schools. It did no more "than to provide a general program to help parents get their children, regardless of their religion, safely and expeditiously to and from accredited schools." To make sure that his ruling did not in any way question or overturn the traditional interpretation of the Establishment Clause, he added: "The First Amendment has erected a wall between church and state. That wall must be kept high and impregnable. We could not approve the slightest breach. New Jersey has not breached it here."
13. In the preceding three paragraphs I have relied heavily on Lloyd P. Jorgenson's two books, *The State and the Non-Public School, 1825–1925*, pp. 136–144, and *The Founding of Public Education in Wisconsin* (Madison: State Historical Society, 1956), pp. 50–110. I should like to acknowledge my indebtedness to him.
14. For more on the Wisconsin High School in the nineteenth century see my *The Once and Future School: Three Hundred and Fifty Years of American Secondary Education* (New York: Routledge, 1996), pp. 79–91.
15. For rural opposition to the schoolmen see Wayne E. Fuller, *The Old Country School* (Chicago: University of Chicago Press, 1982) and, for a different interpretation, Paul Theobald, *Call School: Rural Education in the Midwest to 1918* (Carbondale: Southern Illinois University Press, 1995).
16. On the diversity of private educational institutions see Beadie and Tolley, *Chartered Schools*, my "Diversification in American Higher Education," in *The Transformation of Higher Learning 1860–1930*, ed. Konrad H. Jarausch (Stuttgart: Klett-Cotta, 1982), pp. 196–206, and James McLachlan, *American Boarding Schools: A Historical Study* (New York: Charles Scribner's Sons, 1970).
17. These figures are a bit misleading because before 1890 the reports of the U.S. Commissioner of Education did not include the students enrolled in the few rural public high schools.
18. On the Platteville Story see my *And Sadly Teach: Teacher Education and Professionalization in American Culture* (Madison: University of Wisconsin Press, 1989), pp. 128–135.

19. "Chapter 323," *Laws of Wisconsin . . . in the Year 1875* (Madison: E. B. Bolens, 1875), pp. 623–629.
20. Joseph Schafer, "Genesis of Wisconsin's Free High School System," *Wisconsin Magazine of History* 11 (1926):148.
21. In *Report of the State Superintendent: 1877* (Madison: E.B. Bolens, 1877), p. liii.
22. In *Report of the State Superintendent: 1882* (Madison: David Atwood, 1882), pp. 25–26.
23. In *Biennial Report of the State Superintendent: 1899/1890* (Madison: Democratic Printing Company, 1890), p. 14.
24. "Chapter 493," *Laws of Wisconsin*, 1909.
25. The term "educators in overalls" is Wayne Fuller's. See Chapter 5 in his *The Old Country School.*
26. Fuller, *The Old Country School*, p. 155.
27. Fuller, *The Old Country School*, p. 235, and David R. Reynolds, *There Goes the Neighborhood: Rural School Consolidation at the Grass Roots in Early Twentieth-Century Iowa* (Iowa City: University of Iowa Press, 1999), pp. 85–89.
28. Fuller, *The Old Country School*, p. 245.
29. See Joseph Schafer, *Four Wisconsin Counties* (Madison: State Historical Society, 1927), pp. 235–237.
30. See my *The Once and Future School*, pp. 65–66.
31. David Tyack, *The One Best System: A History of American Urban Education* (Cambridge: Harvard University Press, 1974), p. 127.
32. Tyack, *The One Best System*, p. 176.
33. See William J. Reese, *Power and the Promise of School Reform: Grassroots Movements during the Progressive Era* (Boston: Routledge & Kegan Paul, 1986), pp. 16–18.
34. David Angus and Jeffrey E. Mirel, *The Failed Promise of the American High School, 1890–1995* (New York: Teachers College, Columbia University, 1999), p. 9.
35. See the illuminating essay of Robert H. Wiebe, "The Social Functions of Public Education," *American Quarterly* 21(1969): 147–164.
36. See G. W. F. Hegel, *Grundlinien der Philosophie des Rechts*, ed. Johannes Hoffmeister, 4th ed.(Hamburg: Felix Meiner, 1955), §294, pp. 256–257, and Ursula Krautkrämer, *Staat und Erziehung* (München: Johannes Berchmans Verlag, 1979), pp. 245–249.
37. See Wilhelm von Humboldt, "Der Königsberger und der Litauische Schulplan," in *Schriften zur Politik und zum Bildungswesen*, in *Werke*, ed. Andreas Flitner and Klaus Giel (Stuttgart: J. G. Cotta, 1964), 4: 168–195.

38. See Karl Bungardt, *Die Odysee der Lehrerschaft: Sozialgeschichte eines Standes* (Hannover: Hermann Schrodel, 1965), pp. 29–30.
39. See Anthony J. LaVopa, *Prussian Schoolteachers: Profession and Office, 1763–1848* (Chapel Hill: University of North Carolina Press, 1980), p. 123.
40. See §8 of Süvern's 1819 draft of a general Prussian school law in Lothar Schweim, ed., *Schulreform in Preussen, 1809–1919: Entwürfe und Gutachten* (Weinheim: Julius Beltz, 1966), pp. 128–131.
41. Süvern to Theological Faculty Berlin, December 15, 1823, ms. in Zentrales Staatsarchiv Merseburg, repository 76, VII, new section 1C, general, part 1, no. 1, vol. 1.
42. §87 of Süvern's 1819 draft in Schweim, *Schulreform*, p. 200.
43. §86 of Süvern's 1819 draft in Schweim, *Schulreform*, pp. 197–198.
44. §§19–21 of Süvern's 1819 draft in Schweim, *Schulreform*, pp. 144–146.
45. §87 of Süvern's 1819 draft in Schweim, *Schulreform*, p. 201.
46. For a detailed discussion of Süvern's draft see Luise Wagner-Winterhager, *Schule und Eltern in der Weimarer Republik* (Weinheim und Basel: Beltz Verlag, 1979).
47. Beckedorff's memorandum on Süvern's proposal, submitted in a government bill, is reprinted in Schweim, *Schulreform*, pp. 222–244. The quotations are from pp. 228, 233–234.
48. Beckedorff, Fragment concerning Teacher Seminaries, no. 1820, ms. in Geheimes Staats Archiv Berlin-Dahlem, repository 92, no. 25, pp. 61–64.
49. F. W. III to Altenstein, June 15, 1822, ms. in Zentrales Staats Archiv Merseburg, repository 76, VII, new Section 1C, general, part 1, no. 1, vol. 1, and memorandum of Richter, February 10, 1822; *ibid.*
50. Altenstein to all consistories and provincial governments, July 24, 1822, ms. in Geheimes Staats Archiv Berlin-Dahlem, repository 92, no. 25, pp. 311–314.
51. Karl Friedrich Wilhelm Wander, *Der Kampf um die Schule: Bildungspolitische und Pädagogische Schriften* (Berlin: Volk and Wissen, 1979), 2:41. For more information on the Prussian teachers' role in the revolution of 1848 see Nipperdey, "Volksschule und Revolution im Vormärz," in *Politische Ideologien und Nationalstaatliche Ordnung, Festschrift für Theodor Schieder*, ed. Kurt Kluxen and Wolfgang J. Mommsen (München und Wien: Oldenbourg, 1968), p. 113, and Wilhelm Appens, *Die pädagogischen Bewegungen des Jahres 1848* (Elberfeld: Sam. Lucas, 1917). Cf. also Franzjörg Baumgart, *Zwischen*

Reform und Reaktion: Preußische Schulpolitik 1806–1859 (Darmstadt: Wissenschaftliche Buchgesellschaft, 1900), pp. 144–145, 162.

52. Reprinted in Berthold Michael and Heinz-Hermann Schepp, eds., *Politik und Schule von der Französischen Revolution bis zur Gegenwart; Eine Quellensammlung zum Verhältnis von Gesellschaft, Schule und Staat im 19. und 20. Jahrhundert* (Frankfurt: Athenäum Verlag, 1973), 1:313–314.
53. See Ferdinand Stiehl, *Die drei preussischen Regulative vom 1., 2. und 3. Oktober 1854 über Einrichtung des evangelischen Seminar-, Präparanden- und Elementarschulunterrichts* (Berlin: Hertz, 1854).
54. *Regulativ für den Unterricht in den evangelischen Schullehrer-Seminarien der Monarchie* (Berlin: Schade, 1854), pp. 3–4.
55. "Regulativ für die Einrichtung und Unterricht der evangelischen einklassigen Elementarschule," in Michael and Schepp, *Politik und Schule*, pp. 315–316.
56. Karl-Ernst Jeismann, "Die Stiehlschen Regulative," in *Dauer und Wandel der Geschichte*, ed. Rudolf Vierhaus und Manfred Botzenhart (Münster: Aschendorff, 1966), p. 432.
57. Cf. Peter Martin Roeder, "Gemeindeschule in Staatshand: Zur Schulpolitik des Preußischen Abgeordnetenhauses," *Zeitschrift für Pädagogik* 12 (1966): 560–561.
58. Roeder, "Gemeindeschule in Staatshand," pp. 552–555.
59. Roeder, "Gemeindeschule in Staatshand," pp. 556, 558.
60. Adolph Diesterweg, "Die Freie Schule im Freien Staat," in *Schriften und Reden in zwei Bänden*, ed. Heinrich Deiters (Berlin/Leipzig, Volk und Wissen 1950), 2: 507–511.
61. See on this Luise Wagner-Winterhager, *Schule und Eltern in der Weimarer Republik: Untersuchungen zur Wirksamkeit der Elternbeiräte in Preußen und der Elternräte in Hamburg 1918–1922* (Weinheim und Basel: Beltz Verlag, 1979), pp. 30–32.
62. See Karl-Ernst Jeismann, "Preußische Bildungspolitik vom ausgehenden 18. bis zur Mitte des 19. Jahrhunderts. Thesen und Probleme," in *Zur Bildungs—und Schulgeschichte Preußens*, ed. Udo Arnold (Lüneburg: Nordostdeutsches Kulturwerk, 1988), pp. 29–30.
63. See my *The Once and Future School*, p. 19.
64. Baumgart, *Zwischen Reform und Reaktion*, p. 97.
65. Detlef K. Müller, "The Process of systematisation: The case of German secondary education," in *The Rise of the Modern Educational System: Structural Change and Social Reproduction, 1870–1920*, ed. Detlef K. Müller, Fritz Ringer and Brian Simon (Cambridge: Cambridge University Press, 1987), p. 23.

66. Müller, "The Process of systematisation," p. 24.
67. See documents in Michael and Schepp, *Politik und Schule*, 1: 414–421.
68. Ludwig Schacht, *Über die Gleichberechtigung der Realschule I. Ordnung mit dem Gymnasium* (Elberfeld: Lucas, 1878), p. 66.
69. Hans-Georg Herrlitz, Wulf Hopf, and Hartmut Titze, *Deutsche Schulgeschichte von 1800 bis zur Gegenwart: Eine Einführung* (Weinheim and München: Juventa, 1993), p. 107.
70. Herrlitz, Hopf, and Titze, *Deutsche Schulgeschichte*, p. 109.
71. Michael and Schepp, *Politik und Schule*, 1: 409–414.
72. Georg Kerschensteiner, *Staatsbürgerliche Erziehung der deutschen Jugend: Gekrönte Preisarbeit* (Erfurt: K. Villaret, 1901).
73. James C. Albisetti, *Schooling German Girls and Women: Secondary and Higher Education in the Nineteenth Century* (Princeton, NJ: Princeton University Press, 1988), p. 158.
74. Albisetti, *Schooling German Girls and Women*, pp. 206–209.
75. Michael and Schepp, *Politik und Schule*, 1: 423.
76. Albisetti, *Schooling German Girls and Women*, p. 304.
77. Michael and Schepp, *Politik und Schule*, 1: 425–427.
78. Albisetti, *Schooling German Girls and Women*, p. 293.

3 School Governance and School Choice 1900–1950

1. Raymond E. Callahan reports on Page 214 of his *Education and the Cult of Efficiency: A Study of the Social Forces that have Shaped the Administration of the Public Schools* (Chicago: University of Chicago Press, 1962), that in the academic year 1909–1910 Teachers College awarded 73 graduate degrees in education that included 13 with professional diplomas in administration and supervision. In 1923–1924 that number had grown to 939 graduate degrees with 390 professional diplomas in administration and supervision.
2. Quoted in Callahan, *Education and the Cult of Efficiency*, p. 217.
3. Callahan, *Education and the Cult of Efficiency*, p. 220.
4. On the trend to view teachers as employees rather than professionals see my "Professionalization in Public Education, 1890–1920: The American High School Teacher," in *Bildungsbürgertum im 19. Jahrhundert: Bildungssystem und Professionalisierung im internationalen Vergleich*, ed. Werner Conze and Jürgen Kocka (Stuttgart: Klett-Cotta,

1985), pp. 495–528, and the Chapter on "Professionalization: The Betrayal of the Teacher," in my *And Sadly Teach: Teacher Education and Professionalization in American Culture* (Madison: University of Wisconsin Press, 1989), pp. 161–184.

5. David Tyack and Elisabeth Hansot, *Managers of Virtue: Public School Leadership in America, 1820–1980* (New York: Basic Books, 1951), p. 106.
6. For the school–home relationship in the early twentieth century and the situation in Delaware see William W. Cutler III, *Parents and Schools: The 150 Year Struggle for Control in American Education* (Chicago: University of Chicago Press, 2000), pp. 8, 115–126.
7. On this see William J. Reese, *Power and the Promise of School Reform: Grassroots Movements During the Progressive Era* (Boston: Routledge & Kegan Paul, 1986), pp. 46–50.
8. Reese, *Power and the Promise of School Reform*, p. 79.
9. Quotations are to be found in Lawrence A. Cremin, *The Transformation of the School: Progressivism in American Education, 1876–1957* (New York: Alfred A. Knopf, 1961), pp. 245, 246, 250.
10. Cremin, *The Transformation*, pp. 250–251.
11. Cutler, *Parents and Schools*, pp. 66–67.
12. The address appeared first as "Dare Progressive Education Be Progressive?" *Progressive Education* 9(1932): 257–263.
13. See Chapter 6, "Progressive Education under Fire," in Arthur Zilversmit, *Changing Schools: Progressive Education Theory and Practice, 1930–1960* (Chicago: University of Chicago Press, 1993), pp. 103–117.
14. Cremin, *The Transformation*, p. 327.
15. Cutler, *Parents and Schools*, pp. 134–135.
16. Quoted in Daniel Calhoun, ed., *The Educating of Americans: A Documentary History* (Boston: Houghton Mifflin Company, 1969), p. 42.
17. Reese, *Power and the Promise of School Reform*, p. 162.
18. Cutler, *Parents and Schools*, p. 2.
19. James Bryant Conant, *The Child, The Parent, and the State* (Cambridge: Harvard University Press, 1959), p. 47.
20. Cutler, *Parents and Schools*, p. 44.
21. See my account of these developments in *The Once and Future School: Three Hundred and Fifty Years of Secondary American Education* (New York: Routledge, 1996), pp. 165–185.
22. David L. Angus and Jeffrey E. Mirel, *The Failed Promise of the American High School, 1890–1995* (New York: Teachers College, Columbia University, 1999), p. 65.

23. See Educational Policies Commission, *Education for All American Youth* (Washington, DC: NEA, 1944), pp. 35, 39, 192, 211, 296, 359.
24. James Bryant Conant, *Slums and Suburbs* (New York: McGraw-Hill, 1961), pp. 145–147.
25. See my *The Once and Future School*, p. 188, and James Bryant Conant's *The Comprehensive High School: A Second Report to Interested Citizens* (New York: McGraw-Hill, 1967), pp. 2, 20.
26. Data are taken from various volumes of the *Report of the Commissioner, United States Bureau of Education*, from the U.S. Bureau of the Census, *Historical Statistics of the United States, Colonial Times to 1970* (Washington, DC: U.S. Department of Commerce, 1975), pp. 368–369, Series H 412–432, and the National Center for Education Statistics, *Digest of Education Statistics 1999*, Table 3.
27. For a comprehensive review of Catholic education in the United States see Kim Tolley, " 'Many Years before the Mayflower': Catholic Academies and the Development of Parish High Schools in the United States," in *Chartered Schools: Two Hundred Years of Independent Academies in the United States, 1727–1925*, ed. Nancy Beadie and Kim Tolley (New York and London: RoutledgeFalmer, 2002), pp. 304–328.
28. Otto F. Kraushaar, *American Nonpublic Schools: Patterns of Diversity* (Baltimore: Johns Hopkins University Press, 1972), p. 77.
29. See Lawrence Cremin's engaging description of the progressive schools in his *The Transformation of the School*, pp. 179–239.
30. *Pierce v. Society of the Sisters of the Holy Names of Jesus and Mary*, 268 U.S. 510 (1925).
31. *Farrington v. T. Tokushige*, 273 U.S. 284 (1927).
32. *Cochran v. Louisiana State Board of Education*, 281 U.S. 370 (1930).
33. *West Virginia State Board of Education v. Barnette*, 319 U.S. 624 (1943).
34. *McCollum v. Board of Education*, 333 U.S. 203 (1948) and *Zorach v. Clausen*, 343 U.S. 306 (1952).
35. §143, Weimar Constitution. A convenient selection of the articles on schooling and education is printed in Karl-Heinz Nave, *Die allgemeine deutsche Grundschule: Ihre Entstehung aus der Novemberrevolution von 1918* (Weinheim: Verlag Julius Beltz, 1961), pp. 165–167. The entire constitution may be consulted at <http://www.uni-wuerzburg.de/rechtsphilosophie/wrv1919.html>.
36. §144, Weimar Constitution.
37. §145, Weimar Constitution.
38. §146, Weimar Constitution.

39. §146, Weimar Constitution.
40. §146, Weimar Constitution.
41. Cf. on this Luise Wagner-Winterhager, *Schule und Eltern in der Weimarer Republik: Untersuchungen zur Wirksamkeit der Elternbeiräte in Preußen und der Elternräte in Hamburg 1918–1922* (Weinheim and Basel: Beltz Verlag, 1979), pp. 56–60.
42. §147, Weimar Constitution.
43. §147, Weimar Constitution.
44. See Nave, *Die allgemeine deutsche Grundschule*, p. 63.
45. §147, Weimar Constitution.
46. The *Grundschul* law is reprinted in Christoph Führ, *Zur Schulpolitik der Weimarer Republik*, 2nd ed. (Weinheim: Beltz, 1972), pp. 161–162. The city-state of Hamburg provides a good example of the circumstances under which the private *Vorschulen* continued to exist. See Hildegard Milberg, *Schulpolitik in der pluralistischen Gesellschaft: Die politischen and sozialen Aspekte der Schulreform in Hamburg 1890–1935* (Hamburg: Leibniz Verlag, 1970), pp. 199–206.
47. §120, Weimar Constitution.
48. Führ, *Zur Schulpolitik*, p. 37.
49. Catholic and Protestant interpretations of the *Elternrecht*, however, were not identical. The Catholic view declared the school to be an institution auxiliary to the family whose duty it was to provide their children with a Catholic-confessional education and whose rights at the same time were limited by the church's primary claim to educate its children. For Protestants the right to educate rested not with the church but equally with the parents and the state. The revolution of 1919, however, had brought about the separation of state and church and parents now would claim the right to choose a confessional school for their children. See Milberg, *Schulpolitik in der pluralistischen Gesellschaft*, pp. 156–157.
50. See on this Wagner-Winterhager, *Schule und Eltern*, pp. 66–71.
51. See Nave, *Die allgemeine deutsche Grundschule*, pp. 89–91. It is of interest to note that on Page 90 Karl-Heinz Nave, himself a strong supporter of the common *Grundschule*, points to the "obvious contrast" in which the compulsory attendance at these schools in Germany stands to attendance at elementary schools in other *democratic* countries—he mentions the United States, Great Britain, and France—where compulsory attendance at public elementary schools is a *minimum* requirement and could be fulfilled at private schools as well.

52. The law is reprinted in Führ, *Zur Schulpolitik*, p. 163.
53. See Wagner-Winterhager, *Schule und Eltern*, p. 48.
54. For the situation in Hamburg see Milberg, *Schulpolitik in der pluralistischen Gesellschaft*, pp. 59–62 and 246–258.
55. Cf. Wagner-Winterhager, *Schule und Eltern*, pp. 53–283. Hans-Peter de Lorent and Volker Ullrich, eds., *Der Traum von der freien Schule: Schule und Schulpolitik in der Weimarer Republik* (Hamburg: Ergebnisse Verlag, 1988), presents accounts of the very similar situation in Hamburg.
56. See Jürgen Oelkers, "Origin and Development in Central Europe," in *Progressive Education Across the Continents*, ed. Hermann Röhrs and Volker Lenhart (Frankfurt am Main: Peter Lang, 1995), pp. 31–50.
57. See George Mosse on Kurt Hahn's unwillingness to accept parents as educators in his *Confronting History: A Memoir* (Madison: University of Wisconsin Press, 2000), p. 57.
58. See the 1911 debate between Kerschensteiner and Hugo Gaudig, discussed by Jürgen Oelkers in *Reformpädagogik: Eine Kritische Dogmengeschichte* (Weinheim: Juventa Verlag, 1989), p. 155, and by Hermann Weimer and Juliane Jacobi in *Geschichte der Pädagogik*, 19th ed. (Berlin: Walter de Gruyter, 1991), pp. 188–190.
59. For an evaluation of the Jena plan see Oelkers, *Reformpädagogik*, pp. 158–165.
60. Hanno Schmitt, "Zur Realität der Schulreform in der Weimarer Republik," in *Politische Reformpädagogik*, ed. Tobias Rülcker and Jürgen Oelkers (Bern: Peter Lang, 1998), pp. 619–643.
61. Jürgen Oelkers, "Origin and Development in Central Europe," in *Progressive Education Across the Continents*, ed. Röhrs and Lenhart, p. 33.
62. See Hildegard Feidel-Mertz, ed., *Schulen im Exil: Die verdrängte Pädagogik nach 1933* (Reinbeck bei Hamburg: Rowohlt, 1983), pp. 25–33.
63. Feidel-Mertz, *Schulen im Exil*, pp. 22–24.
64. Feidel-Mertz, *Schulen im Exil*, pp. 33–51, and Hermann Weimer and Juliane Jacobi, *Geschichte der Pädagogik* (Berlin: Walter de Gruyter, 1992), pp. 196–200.
65. See Hans-Georg Herrlitz, Wulf Hopf, and Hartmut Titze, *Deutsche Schulgeschichte von 1800 bis zur Gegenwart: Eine Einführung* (Weinheim and München: Juventa Verlag, 1993), p. 151.
66. Feidel-Mertz, *Schulen im Exil*, pp. 20–21.

4 School Choice in the United States after World War II

1. *Brown v. Board of Education*, 347 U.S. 483 (1954).
2. For summaries of various measures adopted in Southern states to avoid desegregation see Herbert O. Reid, "The Supreme Court Decision and Interposition," *The Journal of Negro Education* 25 (Spring 1956): 109–110, and Walter F. Murphy, "Desegregation in Public Education—A Generation of Future Litigation," *Maryland Law Review* 15: 221–243.
3. These measures were popular in North Carolina, Florida, Alabama, and Mississippi. See Numan V. Bartley, *The Rise of Massive Resistance: Race and Politics in the South During the 1950s* (Baton Rouge: Louisiana State University Press, 1969), pp. 77–78.
4. Bartley, *The Rise of Massive Resistance*, pp. 76–77.
5. For the developments in Alabama see Edward R. Crowther, "Alabama's Fight to Maintain Segregated Schools, 1953–1956," *Alabama Review* 43 (July 1990): 206–225.
6. Congressional Record, 84th Cong., 2nd sess., 1956, 102: 3948, 4004. Cited in Robbins L. Gates, *The Making of Massive Resistance: Virginia's Politics of Public School Desegregation, 1954–1956* (Chapel Hill: University of North Carolina Press, 1962), p. 118.
7. Bartley, *The Rise of Massive Resistance*, pp. 320–325.
8. Gates, *The Making of Massive Resistance*, pp. 117, 123.
9. Gates, *The Making of Massive Resistance*, p. 212.
10. For a detailed description of the developments and issues concerning the "freedom of choice" program in Virginia and Prince Edward County see *Griffin v. County School Board of Prince Edward County*, 377 U.S. 218 (1964) and Amy E. Murrell, "The 'Impossible' Prince Edward Case: The Endurance of Resistance in a Southside County, 1959–64," in *The Moderates' Dilemma: Massive Resistance to School Desegregation in Virginia*, ed. Matthew D. Lassiter and Andrew B. Lewis (Charlottesville: University Press of Virginia, 1998), pp. 134–167.
11. For this see Michael W. Fuquay, "Civil Rights and the Private School Movement in Mississippi, 1964–1971," *History of Education Quarterly* 42 (Summer 2002): 159–180.
12. *Poindexter v. Louisiana Financial Assistance Commission*, 275 F. Supp. 833 (1967).
13. *Green et al. v. County School Board of New Kent County et al.*, 391 U.S. 430 (1968). In *Keyes v. School District No. 1*, 413 U.S. 189 (1973) the

Supreme Court similarly ruled that the use of optional attendance zones was unconstitutional when they created or maintained racially or ethnically segregated schools.

14. Milton Friedman, "The Role of Government in Education," in *Economics and the Public Interest*, ed. Robert A. Solo (New Brunswick, NJ: Rutgers University Press, 1955).
15. A representative example of the reform literature is John C. Holt, *How Children Fail* (New York: Pitman, 1964); on community control see Maurice R. Berube and Marilyn Gittell, eds., *Confrontation at Ocean Hill-Brownsville: The New York School Strikes of 1968* (New York: Praeger, 1969); on homeschooling see J. Gary Knowles, Stacey E. Marlow, and James A. Muchmore, "From Pedagogy to Ideology: Origins and Phases of Home Education in the United State, 1970–1990," *American Journal of Education* 100 (February 1992): 195–235; and on deschooling see Ivan Illich, *Deschooling Society* (New York: Harper and Row, 1971).
16. Milton Friedman, *Capitalism and Freedom* (Chicago: University of Chicago Press, 1962).
17. Theodore Sizer and Phillip Whitten, "A Proposal for a Poor Children's Bill of Rights," *Psychology Today* 2(August, 1968): 59–63, and Theodore Sizer, "The Case for a Free Market," *Saturday Review* 93 (January 1969): 34–42.
18. Christopher Jencks, "Education Vouchers: Giving Parents Money to Pay for Schooling," *The New Republic* 163 (July 4, 1970): 20.
19. Daniel Weiler, *A Public School Voucher Demonstration, the First Year at Alum Rock: Prepared for the National Institute of Education* (Santa Monica, CA: Rand, 1974).
20. Weiler, *A Public School Voucher Demonstration*, p. 6.
21. John E. Coons and Stephen D. Sugarman, *Education by Choice: The Case for Family Control* (Berkeley: University of California Press, 1978). In their *Scholarships for Children* (Berkeley: Institute of Governmental Studies Press, 1992), pp. 3 and 4, Coons and Sugarman summed up their position that any choice system had to "tilt" toward the poor, that it must subsidize choice in both public and private schools, that both public and private schools must be protected from regulation, and that to protect the poor a pure laissez-faire approach on the supply side was unacceptable.
22. *Serrano v. Priest*, 5 Cal. 3d 584 (1971).
23. See Mary Haywood Metz, *Magnet Schools in their Organizational and Political Context* (Washington, DC: National Institute of Education, 1981).

24. Deborah Meier, *The Power of Their Ideas: Lessons for America from a Small School in Harlem* (Boston: Beacon Press, 1995), p. 102.
25. Christine H. Rossell and Charles L. Glenn, "The Cambridge Controlled Choice Plan," *The Urban Review* 20 (Summer 1988): 75–94. The statistics are given on p. 85, and the quotation is taken from p. 92.
26. National Commission on Excellence in Education, *A Nation at Risk: The Imperative for Educational Reform* (Washington, DC: U.S. Government Printing Office, 1983).
27. National Governors' Association, *Time for Results: The Governors' 1991 Report on Education* (Washington, DC: National Governors' Association, 1986), p. 12.
28. Task Force on Teaching as a Profession, *A Nation Prepared: Teachers for the 21st Century* (New York: Carnegie Forum on Education and the Economy, 1986), pp. 14, 92–93.
29. John E. Chubb and Terry Moe, *Politics, Markets, and America's Schools* (Washington, DC: Brookings Institution, 1990), pp. 188, 215, 212.
30. Chubb and Moe, *Politics, Markets, and America's Schools*, pp. 218–219, 221–224.
31. Chubb and Moe, *Politics, Markets, and America's Schools*, p.220.
32. Christian Braunlich and Melanie Looney, eds., *Charter Schools 2002: Results from CER's Annual Survey of America's Charter Schools* (Washington, DC: The Center for Education Reform, 2002), p. 2. The data for 2005 are taken from the CER website.
33. In the 2001–2002 school year, the State of Colorado provided a per pupil capital stipend to charter schools that spent more than 3 percent of their operating budget on capital needs to assist with the purchase, construction, or renovation of their buildings.
34. For more on this see Abby R. Weiss, *Going It Alone: A Study of Massachusetts Charter Schools* (Boston: Northeastern University Institute for Responsive Education, 1997).
35. *The Rocky Mountain News* of October 26, 2002, reports this from Aurora, Boulder Valley, and Fort Collins, Colorado.
36. "Report critical of charter schools," *San Francisco Chronicle*, April 8, 2003.
37. See "Social Change: Charter Schools Take Root," *San Francisco Chronicle*, October 6, 2002.
38. Howard Gardner, "Paroxysms of Choice," *The New York Review of Books*, October 19, 2000, 46–47.

39. Martha Naomi Alt and Katharin Peter, “Private Schools: A Brief Portrait,” *The Condition of Education 2002* (Washington, DC: National Center for Education Statistics, 2002), pp. 5, 9, 17, 19.
40. Alt and Peter, “Private Schools,” p. 21.
41. See the description of John E. Coons and Stephen D. Sugarman, *Scholarships for Children* (Berkeley: Institute of Governmental Studies Press, 1992), pp. 61–68. While by 2004 the Milwaukee voucher program was rocked by corruption charges and on August 19, 2005, the financial officer of a voucher school was found guilty of embezzlement, the number of schools in the program continued to grow. Reported in *The Denver Post* in an Associated Press release of April 6, 2004 and in an editorial of April 12, 2004, and in the *Milwaukee Journal Sentinel* of August 13 and 20, 2005.
42. *Zorach v. Clauson*, 343 U.S. 306, 315 (1952).
43. See chapter 3, note 30.
44. *Wisconsin v. Yoder*, 406 U.S. 205, 217, 234 (1972).
45. *Perchemlides v. Frizzle*, 16641 Massachusetts Hampshire County Superior Court (1978).
46. *Lemon v. Kurtzman*, 403 U.S. 602, 612–613 (1971).
47. *Mueller v. Allen*, 463 U.S. 388 (1983).
48. The First Amendment's establishment clause reads: “Congress shall make no law respecting an establishment of religion, or prohibiting the free exercise thereof . . .”
49. *Zelman v. Simmons-Harris*, 536 U.S. 639 (2002).
50. *Elementary and Secondary Education Act*, Public Law 89–10 (April 11, 1965).
51. Public Law 89–10 (April 11, 1965), quoted in Marvin Lazerson, ed., *American Education in the Twentieth Century: A Documentary History* (New York: Teachers College Press, 1987), pp. 163, 164.
52. *Everson v. Board of Education of Ewing TP*, 330 U.S. 1, 18 (1947).
53. See Nancy Paulu, *Improving Schools and Empowering Parents: Choice in American Education—A Report based on the White House Workshop on Choice in Education* (Washington, DC: U.S. Government Printing Office, 1989). The quotations appear on pp. iii and 32.
54. H.R. 1804, *Goals 2000: Educate America Act* (January 25, 1994).
55. Center for Education Reform, *News Alert: The Bush Administration Education Proposal* (January 23, 2001).
56. As reported by Charles Babington in the *Washington Post* of January 23, 2001.
57. Public Law 107–110.

5 *Schulwahl* in the Post–World War II Period

1. The Potsdam Agreement, published in the Official Papers of the Allied Control Council in Germany, 1946, additional paper no. 1, p. 13.
2. From the Law for the Democratization of the German School, May/June 1946, quoted in Karl-Heinz Günther and Gottfried Uhlig, eds., *Dokumente zur Geschichte des Schulwesens in der Deutschen Demokratischen Republik—Teil 1:1945–1955* (Berlin: Verlag Volk and Wissen, 1970), pp. 207–208.
3. See "OMGUS-Telegramm an die Militärregierungen der vier Länder der amerikanischen Besatzungszone, Januar 10, 1947," in *Bildungspolitik und Bildungsreform*, ed. Leonard Froese and Victor von Blumenthal (München: Wilhelm Goldmann Verlag, 1969), pp. 100–101.
4. See "Direktive Nr. 54 der Allierten Kontrollbehörde in Deutschland, 25. Juni 1947," in Froese und Blumenthal, *Bildungspolitik*, p. 102.
5. See Peter Lundgreen, *Sozialgeschichte der deutschen Schule im Überblick, Teil II: 1918–1980* (Göttingen: Vandenhoeck & Ruprecht, 1981), p. 18.
6. See Froese and Blumenthal, *Bildungspolitik*, pp. 106–113.
7. Froese and Blumenthal, *Bildungspolitik*, p. 48, and Oskar Anweiler, *Schulpolitik and Schulsystem in der DDR* (Opladen: Leske und Budrich, 1988), p. 26.
8. Froese and Blumenthal, *Bildungspolitik*, pp. 36–38, 114–121.
9. Lundgreen, *Sozialgeschichte*, p. 26.
10. See *inter alia* Lutz R. Reuter, "Das Recht auf Bildung in der deutschen Bildungsgeschichte seit 1945," in *50 Jahre Grundgesetz und Schulverfassung: Abhandlungen zu Bildungsforschung und Bildungsrecht*, ed. Frank Rüdiger Jach und Siegfried Jenkner (Berlin: Duncker & Humblot, 2000), 4:2.
11. On this see Christoph Führ, *The German Education System since 1945* (Bonn: Inter Nationes, 1997), p. 10, and Lundgreen, *Sozialgeschichte*, pp. 25–26 .
12. See Günther and Uhlig, *Dokumente—Teil 1: 1945–1955*, p. 208, and *Teil 2: 1956–1967/68, 1st Halbband*, p. 318.
13. Führ, *The German Education System*, p. 13.
14. Froese and Blumenthal, *Bildungspolitik*, pp. 48–50. See also Anweiler, *Schulpolitik*, p. 45.
15. Günther and Uhlig, *Dokumente—Teil 1: 1945–1955*, p. 382.
16. See Bernd Zymek, "Die Schulentwicklung in der DDR im Kontext einer Sozialgeschichte des deutschen Schulsystems," in *Bildungsgeschichte einer*

Diktatur: Bildung und Erziehung in SBZ und DDR im historisch-gesellschaftlichen Kontext, ed. Sonja Häder and Heinz-Elmar Tenorth (Weinheim: Deutscher Studien Verlag, 1997), pp. 36–38.

17. Günther and Uhlig, *Dokumente—Teil 2: 1956–1967/68, 1st Halbband*, pp. 315–323.
18. Cf. Anweiler, *Schulpolitik*, pp. 71–72.
19. Cited in Froese and Blumenthal, *Bildungspolitik*, p. 192.
20. Dietmar Waterkamp, *Handbuch zum Bildungswesen der DDR* (Berlin: Berlin Verlag, 1987), pp. 109–110, and Günther and Uhlig, *Dokumente—Teil 3: 1968–1972/73, 1st Halbband*, pp. 124–127. Technical instruction was added in 1980; see Günther and Uhlig, *Dokumente—Teil 4: 1973–1980/81*, p. 429. For some of the internal debates concerning the 1965 law see Karl-Heinz Günther, *Rückblick* (Frankfurt: Peter Lang, 2002), pp. 381–384.
21. Günther and Uhlig, *Dokumente—Teil 2: 1956–1967/68, 2nd Halbband*, p. 683.
22. Günther and Uhlig, *Dokumente—Teil 2: 1956–1967/68, 2nd Halbband*, p. 685.
23. In addition to the special schools mentioned in the text, there existed schools for children with physical or mental handicaps.
24. Cited in Froese and Blumenthal, *Bildungspolitik*, p. 281.
25. Anweiler, *Schulpolitik*, p. 98.
26. Günther and Uhlig, *Dokumente—Teil 3: 1968–1972/73, 1st Halbband*, pp. 44–53.
27. Günther and Uhlig, *Dokumente—Teil 3: 1968–1972/73, 1st Halbband*, p. 46.
28. Günther and Uhlig, *Dokumente—Teil 3: 1968–1972/73, 2nd Halbband*, pp. 671–676.
29. For the statistics I have relied on Anweiler, *Schulpolitik*, pp. 131–132.
30. One should keep in mind, however, that, as Zymek argued, different local conditions of schooling and the various subdivisions within the *Erweiterte Oberschule*, "while formally constituting an orderly unitary system, nevertheless created different conditions of schooling and social composition which opened up to students differentiated opportunities for learning and employment." Unless families changed their place of residence, parents could not utilize these differences to exercise school choice. See Zymek, "Die Schulentwicklung," p. 43.
31. See article 37 of the October 7, 1949, Constitution of the DDR, the governmental decree about the formation and tasks of parental boards of April 12, 1951, and the similar decree of October 14, 1955, in Günther

and Uhlig, *Dokumente—Teil 1: 1945–1955*, pp. 344, 393–395, 525–527.

32. Günther and Uhlig, *Dokumente—Teil 2: 1956–1967/68, 1st Halbband*, pp. 323–328.
33. Günther and Uhlig, *Dokumente—Teil 2: 1956–1967/68, 2nd Halbband*, p. 783.
34. Günther and Uhlig, *Dokumente—Teil 3: 1968–1972/73, 1st Halbband*, pp. 251–253.
35. It should be noted, however, that few, if any, of the new *Länder* have in fact introduced the *Hauptschule*.
36. Wolfgang Mitter, "Allgemeinbildendes Schulwesen: Grundfragen und Überblick," in *Vergleich von Bildung und Erziehung in der Bundesrepublik Deutschland und in der deutschen Demokratischen Republik*, ed. Oskar Anweiler (Köln: Verlag Wissenschaft und Politik, 1990), p. 174. See also Führ, *German Education System*, pp. 78–81.
37. See the examples I give in chapter 4, such as the citizenship, manual work, and student-initiated projects of the *Arbeitsschule* of Georg Kerschensteiner and Hugo Gaudig, the community-centered and parent-involved activities of the Jena Plan schools of Peter Petersen, and the work of the German Central Institute for Education and Instruction.
38. Manfred Weiss, "Expanding the Third Sector in Education? A Critical View," in *Education Between States, Markets, and Civil Society: Comparative Perspectives*, ed. Heinz-Dieter Meyer and William Lowe Boyd (Mahwah, NJ: Lawrence Erlbaum Associates, 2001), p. 172.
39. See *Grundgesetz*, http://www.bundestag.de/gesetze/gg. For a comparison with the constitution of the Weimar Republic see earlier, chapter 3.
40. Basic Law of the Federal Republic, article 7.
41. Quoted by Andreas Flitner, "Freie Schulen—Ergänzung und Herausforderung des öffentlichen Schulsystems," in Flitner and Hans-Ullrich Gallwas, *Privatschulen und öffentliches Schulsystem* (Köln: Deutscher Instituts-Verlag, 1980), p. 10. On the private schools' subsidiary status as compared to public schools see also Winfried Schlaffke and Reinhold Weiß, eds., *Private Bildung—Herausforderung für das öffentliche Bildungsmonopol* (Köln: Deutscher Institutsverlag, 1996), p. 29.
42. Eckhard K. Deutscher, "Private Schulen in der deutschen Bildungsgeschichte: Ein Beitrag zum Verhältnis von Schule und Staat" (Inauguraldissertation, Universität Frankfurt am Main, 1976), 150–155.

43. Reported in *Die Welt*, April 7, 2003.
44. Bundes Verwaltungs Gericht Aktenzeichen 5 C 70.88—August 13, 1992. See also Udo Dirnaichner, "Übernahme von Schulgeld einer Privatschule im Rahmen der Sozialhilfe," *Schulverwaltung: Zeitschrift für Schulleitung und Schulaufsicht, Ausgabe Bayern* 17 (May 1994): 195–197.
45. Hans Heckel, *Schulrechtskunde: Ein Handbuch für Praxis, Rechtssprechung und Wissenschaft*, 6th ed., ed. Hermann Avenarius and Helmut Fetzer (Berlin: Luchterhand, 1986), pp. 153–155.
46. Max Planck Institut für Bildungsforschung, *Das Bildungswesen in der Bundesrepublic Deutschland: Ein Überblick für Eltern, Lehrer, Schüler* (Hamburg: Rowohlt, 1990), p. 140.
47. See Klaus Hurrelmann, "Tendenzen der Privatisierung des deutschen Schulsystems—Chance oder Gefahr für Qualität und Chancengleichheit der Bildung?" *Forum E: Zeitschrift des Verbandes Bildung und Erziehung* (November/December 1995), pp. 24–26.
48. According to a press statement released by Cardinal Karl Lehmann in March 2003 applications for admission to Catholic schools exceed the available places by 30 percent. Catholic schools enroll nearly 370,000 students, more than half of all the pupils in private schools. Protestant schools report the same discrepancy between the number of applicants and available places. See *Die Welt*, April 7, 2003.
49. Doris Knab and Felix Messerschmid, "Tradition und Gegenwart. Profile von Schulen in katholischer Trägerschaft," in *Alternative Schulen. Gestalt und Funktion nichtstaatlicher Schulen im Rahmen öffentlicher Bildungssysteme*, ed. Dietrich Goldschmidt und Peter Martin Roeder (Stuttgart: Klett-Cotta, 1979), pp. 362–378.
50. Karl-Heinz Potthast, "Evangelische Schulen und Ausbildungsstätten zwischen den Bildungssynoden 1971 und 1978," in *Alternative Schulen*, ed. Goldschmidt and Roeder, pp. 353–361.
51. Flitner and Gallwas, *Privatschulen und öffentliches Schulsystem*, pp. 10–17, 203–205.
52. See Informationsunterlage des Sekretariats der Kultusministerkonferenz, *Übergang von der Grundschule in Schulen des Sekundarbereichs I* (Sekretariat der Ständigen Konferenz der Kultusminister der Länder in der Bundesrepublik Deutschland, June 1999—mimeographed. For a more detailed description of parental involvement see *The Educational System in Germany: Case Study Findings: The Transition to Lower Secondary Education*, June 1999, on the U.S. Department of Education website, http://www.ed.gov/pubs/GermanCaseStudy/chapter2a.html
53. Informationsunterlage, *Übergang*, pp. 4, 6, 7.

54. See on these matters Werner Heldmann, *Kultureller und gesellschaftlicher Auftrag von Schule: Bildungstheoretische Studie zum Schulkonzept* "Die Soziale Leistungsschule" des Philoogen-Verbandes Nordrhein-Westfalen (Krefeld: Pädagogik und Hochschul Verlag, 1990), pp. 300–301, and Christian Starck, *Staat, Schule, Kirche in der Bundesrepublik Deutschland und in Frankreich* (Kehl, Straßburg: N. P. Engel Verlag, 1982), p. 17. For a comprehensive overview of regional differences see "The Educational System in Germany: Case Study Findings, June 1999, chapter 2, The Transition to Lower Secondary Education," at http://www.ed.gov/pubs/GermanCaseStudy/chapter2a.html
55. Informationsunterlage, *Übergang*, pp. 11 and 12.
56. Informationsunterlage, *Übergang*, pp. 12 and 13.
57. Informationsunterlage, *Übergang*, p. 16.
58. See on this Werner Heldmann, *Kultureller und gesellschaftlicher Auftrag von Schule*, pp. 300–309.
59. Jürgen Baumert et al., eds., *PISA 2000: Basiskompetenzen von Schülerinnen und Schülern im internationalen Vergleich* (Opladen: Leske und Budrich, 2001), pp. 430, 431.
60. For a contemporary account see Isabelle de Pommereau, "Germany: Schools that divide," *Christian Science Monitor*, October 22, 2002.
61. Martin Spiewak, "Die Schule brennt," *Die Zeit*, Politik 50/2001.
62. With the OECD median value being 500, the German values were 484 for reading, 490 for mathematics, and 487 for natural science. The corresponding values for the United States were 504, 493, and 499. Baumert et al., *PISA 2000*, pp. 107, 110, 173, and 229.
63. Baumert et al., *PISA 2000*, pp. 103, 105, 108.
64. Baumert et al., *PISA 2000*, p. 356.
65. Baumert et al., *PISA 2000*, p. 373.
66. Baumert et al., *PISA 2000*, p. 375.
67. Baumert et al., *PISA 2000*, pp. 379, 401.
68. Führ, *German Education System*, pp. 50–51. These councils are controversial, and the Christian Democrats have called for their abolition. See http:// www.ed.gov/pubs/GermanCase Study/chapter2a.html
69. The thesis of state domination of Germany's schools has been well stated by Heinz-Elmar Tenorth in Häder and Tenorth, *Bildungsgeschichte einer Diktatur*, p. 75.
70. Frank-Rüdiger Jach, *Schulvielfalt als Verfassungsgebot* (Berlin: Duncker & Humblot, 1991), pp. 9–10.

6 Retrospect and Outlook

1. On this see Everett Webber, *Escape to Utopia: The Communal Movement in America* (New York: Hastings House, 1959) and Frank E. Manuel, ed., *Utopias and Utopian Thought* (Boston: Houghton Mifflin, 1966).
2. See, for example, Ivan Illich, *Deschooling Society* (New York: Harper and Row, 1970).
3. Mark Lilla, "A Tale of Two Reactions," *New York Review of Books*, May 14, 1998, 7.
4. Raymond E. Callahan, *Education and the Cult of Efficiency* (Chicago: University of Chicago Press,1962), and David B. Tyack, *The One Best System: A History of American Urban Education* (Cambridge: Harvard University Press, 1974).
5. See Richard Just, "Voucher Nation? Why School Choice Could Demolish National Unity," *American Prospect Online*, July 11, 2002.
6. Andy Green, *Education and State Formation: The Rise of Education Systems in England, France and the USA* (New York: St. Martin's Press, 1990), pp. 76–81, 308–316.
7. Jürgen Oelkers, *Reformpädagogik: Eine kritische Dogmengeschichte* (Weinheim and München: Juventa Verlag, 1989), pp. 13 and 27.
8. Grundgesetz für die Bundesrepublik Deutschland, article 7, §§2 and 3.
9. Peter W. Cookson, Jr., *School Choice: The Struggle for the Soul of American Education* (New Haven: Yale University Press, 1994), pp. 127–128.
10. John Chubb and Terry Moe, *Politics, Markets, and America's Schools* (Washington, DC: Brookings Institution, 1990), pp. 218–219.
11. John E. Coons and Stephen D. Sugarman, *Education by Choice: The Case for Family Control* (Berkeley: University of California Press, 1978), p. 153.
12. Milton and Rose Friedman, *Free to Choose, A Personal Statement* (New York: Harcourt Brace Jovanovich, 1979), pp. 167, 171.
13. For theoretical justifications of school choice see Chubb and Moe, *Politics, Markets, and America's Schools*; Coons and Sugarman, *Education by Choice: The Case for Family Control*; and Milton Friedman and Rose Friedman, *Free to Choose.*
14. See on this Joseph Kahne, *Reframing Educational Policy: Democracy, Community, and the Individual* (New York: Teachers College Press, 1996).

15. See Nancy Beadie and Kim Tolley, eds., *Chartered Schools: Two Hundred Years of Independent Academies in the United States, 1727–1925* (New York and London: RoutledgeFalmer, 2002).
16. The Tenth Amendment reads: "The powers not delegated to the United States by the Constitution, nor prohibited by it to the States, are reserved to the States respectively, or to the people."
17. See Winfried Schlaffke and Reinhold Weiß, eds., *Private Bildung—Herausforderung für das öffentliche Bildungsmonopol* (Köln: Deutscher Institutsverlag, 1996), pp. 87–88.
18. Dale D. and Bonnie Johnson, *High Stakes: Children, Testing and Failure in American Schools* (Lanham, MD: Rowman and Littlefield, 2002), p. 203.
19. Richard Elmore, "Unwarranted Intrusion," Education Next (Spring 2002), http://educationnext.org/20021/30.html
20. Quoted by Charles Babington in the Washington Post, January 23, 2001.
21. *Zelman v. Simmons-Harris*, 536 U.S. 639 (2002).
22. *Everson v. Board of Education*, 330 U.S. 1, 18 (1947).
23. *Board of Education of Central School District No. 1 v. Allen*, 392 U.S. 236 (1968); *Lemon v. Kurtzman*, 403 U.S. 602 (1971); *Levitt v. Committee for Public Education & Religious Liberty*, 413 U.S. 472 (1973); *Committee for Public Education and Religious Liberty v. Nyquist*, 413 U.S. 756 (1973); *Meek v. Pittenger*, 421 U.S. 349 (1975); and *Wolman v. Walter*, 433 U.S. 229 (1977).
24. *Mueller v. Allen*, 463 U.S. 388, 398,400 (1983).

Name Index

Abbott, Frank C., 165n
Adams, John, 8
Adler, Felix, 77
Albisetti, James C., 173n
Alt, Martha Naomi, 181n
Altenstein, Baron Karl von, 49, 51, 171n
Alvarado, Anthony, 103
Angus, David, 71, 170n, 174n
Anweiler, Oskar, 182n, 183n, 184n
Appens, Wilhelm, 171n

Bailyn, Bernard, 3, 164n
Barnard, Henry, 14–15, 31, 33, 165n
Bartley, Numan V., 178n
Baumert, Jürgen, 186n
Baumgart, Franzjörg, 166n, 167n, 171n, 172n
Beadie, Nancy, 164n, 168n, 169n, 175n, 188n
Beckedorff, Ludolf von, 50–51, 56, 171n
Becker, Hellmut, 5, 164n
Berg, Christa, 167n, 168n
Bergius, Johann Heinrich, 23
Berube, Maurice R., 179n
Black, Justice Hugo, 159, 169n
Blaine, James G., 36
Blair, Henry William, 36
Blumenthal, Victor von, 182n
Bobbitt, John Franklin, 64
Braunlich, Christian, 180n
Breyer, Justice Stephen G., 162
Bungardt, Karl, 171n
Burrow, J. W., 167n
Bush, President George Herbert Walker, 118
Bush, President George W., 120
Byrd, Senator Harry F., 97

Callahan, Raymond, 63–64, 146, 173n, 187n
Cavazos, Lauro F., 118
Chancellor, William Eastbrook, 64
Chubb, John, 105–108, 150, 163n, 180n, 187n
Claparède, Edouard, 87
Clinton, President William, 119
Cobb, Stanwood, 67
Conant, James Bryant, 69–72, 164n, 174n, 175n
Cookson, Peter, 150, 163n, 187n
Coons, John E., 102, 150, 179n, 181n, 187n
Coram, Robert, 9, 22
Counts, George, 67
Cousin, Victor, 6
Cremin, Lawrence A., 3, 4, 68, 77, 164n, 165n, 174n, 175n
Cubberley, Elwood P., 64
Curti, Merle, 4

Cutler, William W., 69, 70, 163n, 174n

de Lorent, Hans-Peter, 177n
Deutscher, Eckhard K., 184n
Diesterweg, Adolph, 54–55, 86, 87, 172n
DuBois, W. E. B., 34, 168n
du Pont de Nemours, Pierre Samuel, 9, 164n

Elliott, Edward C., 64
Elmore, Richard, 158, 188n

Feidel-Mertz, Hildegard, 177n
Fichte, Johann Gottlieb, 27–28, 167n
Flitner, Andreas, 184n, 185n
Friedman, Milton, 99, 100, 150, 179n, 187n
Friedman, Rose, 150, 187n
Friedrich Wilhelm I, King of Prussia, 20
Friedrich Wilhelm III, King of Prussia, 48, 51, 171n
Froese, Leonhard, 166n, 182n
Führ, Christoph, 176n, 177n, 182n, 184n, 186n
Fuller, Wayne, 42, 164n, 169n, 170n

Gallwas, Hans-Ullrich, 184n, 185n
Gardner Howard, 109, 180n
Gates, Robbins L., 178n
Gaudig, Hugo, 90, 177n, 184n
Geheeb, Edith, 87
Geheeb, Paul, 87, 92
Gittell, Marilyn, 179n
Goldschmidt, Dietrich, 185n
Goslin, Willard, 67
Graham, Robert, 41
Grant, Ulysses, 34
Green, Andy, 187n
Günther, Karl-Heinz, 182n–184n

Häder, Sonja, 183n, 186n
Hahn, Kurt, 88, 92, 177n
Hansot, Elisabeth, 65, 166n, 174n
Harlan, Louis R., 33, 34, 168n
Heckel, Hans, 166n, 185n
Hegel, Georg Wilhelm Friedrich, 47, 170n
Heinemann, Manfred, 167n
Heldmann, Werner, 186n
Herbart, Johann Friedrich, 28, 29, 89, 152, 168n
Herrlitz, Hans-Georg, 173n, 177n
Hitler, Adolf, 92
Hoar, George F., 35
Holt, John C., 179n
Hopf, Wulf, 173n, 177n
Hughes, John, Bishop, 17–19, 166n
Humboldt, Wilhelm von, 7, 25–27, 28, 29, 48, 89, 152, 167n, 170n

Illich, Ivan, 179n, 187n

Jach, Frank Rüdiger, 5, 142, 163n, 164n, 182n, 186n
Jacobi, Juliane, 177n
Jefferson, Thomas, 8–9, 10
Jeismann, Karl-Ernst, 172n
Jencks, Christopher, 100, 102, 179n
Jenkner, Siegfried, 182n
Johnson, Bonnie, 157–158, 188n
Johnson, Dale D., 157–158, 188n
Johnson, President Lyndon B., 117
Jones, Jacqueline, 168n
Jorgenson, Lloyd P., 168n, 169n

Just, Richard, 187n
Justi, Johann Heinrich Gottlob von, 23

Kaestle, Carl F., 15, 165n, 166n
Kahne, Joseph, 187n
Kerschensteiner, Georg, 6, 60, 90, 173n, 177n, 184n
Key, Ellen, 87
Knox, Samuel, 9–10, 22
Kraushaar, Otto F., 77, 163n, 169n, 175n
Krautkrämer, Ursula, 167n, 170n
Kuhlemann, Frank-Michael, 166n, 167n, 168n

Lamm, Richard D., 105
Lange, Helene, 61
Lannie, Vincent, 166n
Lassiter, Matthew D., 178n
LaVopa, Anthony J., 171n
Lehmann, Cardinal Carl, 185n
Lenhart, Volker, 164n, 177n
Lewin, Heinrich, 166n
Lewis, Andrew B., 178n
Lichtwark, Alfred, 87
Lietz, Hermann, 87
Lilge, Frederic, 167n
Lilla, Mark, 145, 187n
Looney, Melanie, 180n
Lundgreen, Peter, 182n
Luserke, Martin, 88

Mann, Horace, 14–15, 31
Manuel, Frank E., 187n
Mayhew, Ira, 13
McLachlan, James, 169n
McQuaid, Bishop Bernard J., 46
Meier, Deborah, 103, 180n
Metz, Mary Haywood, 179n
Michael, Berthold, 172n, 173n
Milberg, Hildegard, 176n, 177n
Mill, John Stuart, 7
Mirel, Jeffrey E., 71, 170n, 174n
Moe, Terry, 105–108, 150, 163n, 180n, 187n
Montessori, Maria, 87, 89
Mosse, George, 177n
Müller, Detlef K., 172n, 173n

Natorp, C. Ludwig, 26
Nave, Karl-Heinz, 175n, 176n

Oelkers, Jürgen, 92, 147–148, 164n, 177n, 187n
Otto, Berthold, 89

Paine, Thomas, 7
Paulu, Nancy, 181n
Pestalozzi, Johann Heinrich, 48
Peter, Katharin, 181n
Petersen, Peter, 90–91, 184n
Pickard, Josiah, 40, 41

Ravitch, Diane, 166n
Reese, William J., 69, 164n, 168n, 170n, 174n
Rehnquist, Chief Justice William H., 160–161
Reynolds, David R., 170n
Ringer, Fritz, 172n
Roeder, Peter Martin, 172n, 185n
Röhrs, Hermann, 164n, 177n
Rudolph, Frederick, 165n
Rush, Benjamin, 9, 22

Schacht, Ludwig, 13
Schafer, Joseph, 176n
Schepp, Heinz-Hermann, 172n, 173n

Schlaffke, Winfried, 184n, 188n
Schleiermacher, Friedrich, 28, 29, 152, 168n
Schmitt, Hanno, 91
Schultz, Stanley K., 168n
Schweim, Lothar, 171n
Scurla, Herbert, 167n
Searing, Edward, 40–41
Seward, William H., 17
Simon, Brian, 172n
Sizer, Theodore, 100, 102, 179n
Smith, Adam, 7
Smith, Samuel, 9–10, 22
Spalding, Bishop John Lancaster, 4
Spaulding, Frank, 64
Spoerr, Kathrin, 133
Starck, Christian, 186n
Steiner, Rudolf, 77, 89, 135
Stevens, John Paul, 162n
Stiehl, Ferdinand, 172n
Strayer, George D., 64
Sugarman, Stephen D., 102, 150, 179n, 181n, 187n
Süvern, Johann Wilhelm von, 49–50, 51, 85–87, 171n

Tenorth, Heinz-Elmar, 183n, 186n
Theobald, Paul, 169n
Titze, Hartmut, 173n, 177n
Tolley, Kim, 164n, 168n, 169n, 175n, 188n
Trapp, Ernst Christian, 25, 28
Tyack, David, 14, 18, 42, 45, 65, 146, 165n, 166n, 170n, 174n, 187n

Uffrecht, Bernhard, 88
Uhlig, Gottfried, 182n–184n
Ullrich, Volker, 177n

Villaume, Peter, 24–25, 27
Vinovskis, Maris A., 15, 16, 165n, 166n

Wagner-Winterhager, Luise, 171n, 172n, 176n, 177n
Wander, Karl Friedrich Wilhelm, 51–52, 171n
Waterkamp, Dietmar, 183n
Webber, Everett, 187n
Webster, Noah, 9, 22
Weiler, Daniel, 179n
Weimer, Hermann, 177n
Weiss, Abby R., 180n
Weiss, Manfred, 184n
Weiß, Reinhold, 184n, 188n
West, E. G., 165n
Wilhelm II, Emperor, 58–59
Williams, Polly, 114
Wilson, Henry, 35
Wöllner, Johann Christoph von, 25
Woodson, Carter Godwin, 34, 168n
Wyneken, Gustav, 87

Zedlitz, Karl Abraham Freiherr von, 23–24, 29
Zilversmit, Arthur, 174n
Zincke, Georg Heinrich, 23n
Zymek, Bernd, 182n, 183n

Subject Index

Abuse of Learning, The
 see under Lilge, Frederic
Addresses to the German Nation
 see under Fichte, Johann Gottlieb
Afro-Americans,
 schools run by, 34
Allgemeine deutsche Grundschule, Die
 see under Nave, Karl-Heinz
Alternative Schulen
 see under Goldschmidt, Dietrich *and* Roeder, Peter Martin
Alum Rock School District, San Jose, California, 101–102
American Association of School Administrators, 70
American Boarding Schools: A Historical Study
 see under McLachlan, James
American Education: The National Experience 1783–1876
 see under Cremin, Lawrence A.
American Missionary Association, 33–34
American Nonpublic Schools
 see under Kraushaar, Otto F.
American Philosophical Society, 9–10
American University in the Province of New York, 11
Americanization programs, 68
Arbeitsschule
 see under Kerschensteiner, Georg

Beamten (civil servants)
 see under Germany *and* Prussia
Bennett Law
 see under schooling in state of Wisconsin
Berlinische Monatsschrift, 167n
 see under Humboldt, Wilhelm von *and* Zedlitz, Karl Abraham Freiherr von
Besitzbürgertum (propertied bourgeoisie), 24, 58, 59
Beverly, Massachusetts
 see under schooling in state of Massachusetts
Bible reading in American public schools, 18, 43, 44
Bildungsbürgertum (educated bourgeoisie), 24, 58, 59
Bildungsgeschichte einer Diktatur
 see under Häder, Sonja *and* Tenorth, Heinz-Elmar
Bildungspolitik und Bildungsreform
 see under Froese, Leonard *and* Blumenthal, Victor von
Bildungswesen in der Bundesrepublik Deutschland, Das, 164
 see under Max Planck Institute for Education Research

Black Reconstruction in America
see under DuBois, W. E. B.
Board of Commissioners of Common Schools
see under schooling in New York City

Call School: Rural Education in the Midwest to 1918
see under Theobald, Paul
Cambridge, Massachusetts, 180n
see also under schooling in state of Massachusetts
Caputh Jewish School, 93
Cardinal Principles of Secondary Education, 47, 68–69
Carnegie Forum on Education and the Economy, 1986, 105, 118
Catholepistemiad,
see under schooling in state of Georgia
Catholic schools, American
establishment of parochial school system, 17–18, 43, 152, 155
frequency of, 75–77, 112–114
neglect of in educational history, 4
problems of, 35–36, 42–44, 46
see also under schooling in New York City
Catholic schools, German, 134–135, 155, 185n
Center for Education Reform (CER), 107, 181n
centralization of public education
see under systematization of American and German public schools
Century of the Child
see under Key, Ellen
Changing Schools: Progressive Education Theory and Practice, 1930–1960
see under Zilversmit, Arthur
Charlottesville, Virginia, Education Summit of 1989, 119
charter schools, 4, 105–109, 147, 152
Charter Schools 2002
see under Braunlich, Christian *and* Looney, Melanie
Chartered Schools: Two Hundred Years of Independent Academies in the United States, 164, 168
see also under Beadie, Nancy *and* Tolley, Kim
child benefit theory, 78, 117
Civilian Conservation Corps, 71
Committee of Ten Report, 46
Committee of Twelve, 65
community control, 4, 17, 19, 100, 103, 152, 179n
see also under schooling in New York City
communitarianism, 152
Comprehensive High School, The
see under Conant, James Bryant
compulsory school attendance laws, 47, 78
Condition of Education 2002
see under Alt, Martha Naomi *and* Peter, Katharin
Confrontation at Ocean Hill-Brownsville,
see under Berube, Maurice R. *and* Gittell, Marilyn *and also under* school choice in New York City
Confronting History: A Memoir
see under Mosse, George

countercultural revolution, 144
country boarding schools
see under Landerziehungsheime Ilsenburg
county superintendents, 40–42
cult of efficiency
see under Callahan, Raymond
Culture Factory, The
see under Schultz, Stanley K.

deschooling, mentioned, 100
Deschooling Society
see under Illich, Ivan
Deutsche Schulgeschichte von 1800 bis zur Gegenwart
see under Herrlitz, Hans-Georg *and* Hopf, Wulf *and* Titze, Hartmut
Deutsche Schulgesetzgebung 1763–1952
see under Froese, Leonhard
Dokumente zur Geschichte des Schulwesens in der Deutschen Demokratischen Republik
see under Günter, Karl-Heinz and Uhlig, Gottfried
drei preussischen Regulative, Die
see under Stiehl, Ferdinand

Education for all American Youth
see under Educational Policies Commission
Education by Choice: The Case for Family Control
see under Coons, John E. *and* Sugarman, Stephen D.
Education and the Cult of Efficiency
see under Callahan, Raymond
Education in the Forming of American Society, 164n
see also under Bailyn, Bernard
Education for all Handicapped Children Act of 1975, 3
Education Management Organizations, 108
Education of the Negro prior to 1861, The
see under Woodson, Carter Goodwin
éducation nouvelle
see under progressive education
Education and Social Change in Nineteenth Century Massachusetts
see under Kaestle, Carl F. *and* Vinovskis, Maris A.
Education and the State
see under West, E. G.
Education and State Formation
see under Green, Andy
educational administration, departments of, 64
Educational Policies Commission, 70–71, 80, 175n
Einheitsschule, 121, 123, 124, 127, 129, 130
see also under Germany, schooling in the German Democratic Republic,
Elementary School Law of 1925 (Germany), 85
Elementary and Secondary Education Act of 1965, 117, 181
Elementary and Secondary Education Act of 1994, 119

Elternrecht (parental rights), 84–85, 87n, 176n
Erweiterte Oberschule, 125–127, 183n
see also under Germany, schooling in the Soviet Occupied Zone and the German Democratic Republic,
Escape to Utopia
see under Webber, Everett
Essays on Education in the Early Republic
see under Rudolph, Frederick
establishment clause
see under U.S. Constitution
Ethical Culture, Society for, 77
Evolution of an Urban School System, The
see under Kaestle, Carl F.

Failed Promise of the American High School, The
see under Angus, David *and* Mirel, Jeffrey E.
Federal Association of German Private Schools, 135
federal government, initiatives of, 33, 95, 100, 117–120,152, 156–162
first amendment
see under U.S. Constitution
Förderstufe, 131, 138
see also under Germany, schooling in the Federal Republic
foreign languages, use of in American schools, 43–44, 46
Four Wisconsin Counties
see under Schafer, Joseph
free enterprise economists, 150
see also Friedman, Milton *and* Friedman, Rose
free exercise clause
see under U.S. Constitution
Free German Youth, 128
Free High School Law,
see under schooling in state of Wisconsin
free market school choice plans, 99–102
Free School Community Wickersdorf
see under Wyneken, Gustav
Free School Society of New York,
see under schooling in New York City
Free School and Work Community
see under Uffrecht, Bernhand
Freedman's Bureau, 33–34
freedom of choice programs
see under southern opposition to racial integration
Fünfzig Jahre Grundgeetz und Schulverfassung
see under Jach, Frank Rüdiger *and* Jenkner, Siegfried

General Land Law of Prussia (1794), 20, 21, 27, 29, 154
German Central Institute for Education and Instruction, 91, 184n
German Education System since 1945, The
see under Führ, Christoph
Germany
abolition of monopoly of *Gymnasium*, 59
abolition of school fees, 59

Beamten, role of in, 58, 59, 64, 81, 92
dominance of public school, 5
dual system of academic and vocational education, 130
Educational System in Germany: Case Study Findings, The, 185: *see also under* Schulwahl
fear of academic proletariat, 58
Jewish students, treatment during Third Reich, 93
private schools closed in Third Reich, 92
public schools as confessional schools, 82–84
private schools forbidden in German Democratic Republic, 124
reeducation policies after 1945, 126
restorative aspects of post-war school reforms, 124, 131
school wars, 58–59
schooling in: Berlin, 123, 137; the Federal Republic, 123, 130–142; the Second Reich, 56–62; the Soviet Occupied Zone and the German Democratic Republic, 124–139; the Third Reich, 92–94, 156; the Weimar Republic, 80–92
schooling and ethnicity, 138–142
schools for girls, 60–62
social democracy and the schools, 60, 85
social reproduction through public schooling, 57
state control of education, 131, 133, 153, 154
three-pronged system of elementary, middle, and higherschools, 123, 131, 135, 137, 138, 140
vertically differentiated secondary school system, 57–58
women's access to higher education, 61–62

Geschichte der Entwicklung der deutschen Volksschule
see under Lewin, Heinrich
Geschichte der Pädagogik
see under Weimer, Hermann *and* Jacobi, Juliane
Goals 2000: Educate America Act, 119, 181n
Going It Alone: A Study of Massachusetts Charter Schools
see under Weiss, Abby R.
Government Policy and Higher Education: A Study of the Regents of the University of the State of New York
see under Abbott, Frank C.
Great School Wars: New York City 1805–1973
see under Ravitch, Diane *and* schooling in New York City
Grundgesetz (Basic Law), 131, 132, 136, 141, 184n, 187n
see also Germany, schooling in the Federal Republic
Grundlinien der Philosophie des Rechts
see under Hegel, Georg Wilhelm Friedrich

Handbuch zum Bildungswesen der DDR
see under Waterkamp, Dietmar
High Stakes: Children, Testing and Failure in American Schools
see under Johnson, Bonnie *and* Johnson, Dale D
Hitler Youth, 94
home schooling, mentioned, 4, 100, 115–116, 179n
see also under *Perchemlides v. Frizzle* (1978)
Home Teacher School
see under Otto, Berthold
house fathers
see under Prussia, school bill of 1819 *and* Süvern Johann Wilhelm von
How Children Fail
see under Holt, John C.

Ideas for an Attempt to Determine the Limits of the State's Effectiveness, 25, 167n
see also under Humboldt, Wilhelm von
Imperial School Conference of June 1900, 58–59
see also under Germany, schooling in the Second Reich
Improving Schools and Empowering Parents
see under Paulu, Nancy
Informationsunterlage des Sekretariats der Kultusministerkonferenz, 185n
see also under Schulwahl
interposition
see under southern opposition to racial integration

Jena Plan, 177n, 184n
see also under Petersen, Peter

Kampf um die Schule, Der
see under Wander, Karl Friedrich Wilhelm
Karlsbad Decrees, 48
Kennedy School, Berlin, 135
Königsberg and Lithuania, plans for education in, 27, 167n, 170n
see also under Humboldt, Wilhelm von
Kultureller und gesellschaftlicher Auftrag von Schule
see under Heldmann, Werner *and Schulwahl*

Landerziehungsheim Ilsenburg
see under Lietz, Hermann
Landerziehungsheime (country boarding schools), 87–89, 92, 135
Law for the Democratization of the German School (1946), 121, 123, 124, 182n
see also under Germany, schooling in the Soviet Occupied Zone and the German Democratic Republic
Law for the Socialist Development of the Schools of the German Democratic Republic (1959), 125
see also under Germany, schooling in the Soviet Occupied Zone and the German Democratic Republic

Law for the Unified Socialist System of Education (1965), 126
see also under Germany, schooling in the Soviet Occupied Zone and the German Democratic Republic
Limits of State Action, The
see under Burrow, J. W. *and* Humboldt, Wilhelm von
Lionel Wilson College Preparatory Academy, Oakland, California, 108–109
local control of schools in the United States, 10
business men, editors, and professionals as supporters of, 32
defended in Wisconsin, 42, 44
diminishing role of, 16, 31, 44
as reason for school choice, 152
Lutheran schools, American, problems of, 43–44

Maclay bill, 17–18
see also under schooling in New York City
magnet schools, 3, 4, 96, 103, 105, 114, 152, 179n
Magnet Schools in their Organizational and Political Context
see under Metz, Mary Haywood
Making of Massive Resistance, The
see under Bartley, Numan V.
Managers of Virtue
see under Tyack, David *and* Hansot, Elisabeth
market forces in education, 4
see also under free market school choice plans
Massachusetts education law of 1789, 10
Massachusetts State Board of Education, 14, 35
massive resistance
see under southern opposition to racial integration
Max Planck Institut für Bildungsgeschichte (Max Planck Institute for Education Research), 133, 164n, 185n
Mis-Education of the Negro, The
see under Woodson, Carter Goodwin
Moderates' Dilemma, The
see under Lassiter, Matthew D. and Lewis, Andrew B.
Modernisierung und Disziplinierung: Sozialgeschichte des deutschen Volksschulwesens 1794–1872
see under Kuhlemann, Frank-Michael
Montessori schools, 135, 156
see also under Montessori, Maria
mothers' clubs, 46, 66

A Nation at Risk, 104, 118, 180n
A Nation Prepared, 105, 180n
National Center for Education Statistics, 175n
National Commission on Excellence in Education
see under A Nation at Risk
National Congress of Mothers, 46
National Council on Education Standards and Testing, 119
National Defense Education Act of 1958, 118

national education
  American proposals for, 9–10, 155
  Fichte's thoughts on, 27–28
  Humboldt's thoughts on, 26
  Schleiermacher's thoughts on, 29
National Education Association, 46–47, 65, 68–69, 70, 80
National Education in the United States of America
  see under du Pont de Nemours, Pierre Samuel
National Governors' Association, 180n
  see also under *Time for Results*
National Governors' Conference of 1986, 104–105, 118
National Institute of Education, 101
national system of education
  attempt to establish in the United States, 9–10, 34–36, 72
National Youth Administration, 71
new education
  see under Progressive education reforms
News Alert: The Bush Administration Education Proposal (2001)
  see under Center for Education Reform
No Child Left Behind Act, 181n
normal schools, mentioned, 14, 16, 37, 40

Ober-Schulkollegium (National Higher School Board), 24, 29
  see also under Zedlitz, Karl Abraham Freiherr von
Ocean Hill Brownsville
  see under schooling in New York City
Odenwaldschule
  see under Geheeb, Edith *and* Geheeb, Paul
Odysee der Lehrerschaft, The
  see under Bungardt, Karl
OECD, 163n, 186n
  see under PISA tests
Old Country School, The
  see also under Fuller, Wayne
On Liberty
  see under Mill, John Stuart
One Best System, The
  see under Tyack, David B.
Origins of Public High Schools, The
  see under Vinovskis, Maris A.

Pädagogischen Bewegungen des Jahres 1848, Die
  see under Appens, Wilhelm
parens patriae, state as, 143
parent advisory boards in the DDR, 129–130
parent councils in the Weimar Republic, 86–87
parent-school relationship in the United States, 69–70
Parent-Teacher Organizations in the United States, 46, 65–66, 70
parents
  as supporters of German *Schullandheime*, 92
  as school board members in Prussia, 55
  rights of, 25, 80, 84, 87, 141–142, 144
  roles played in American schools, 13–16, 18, 31, 51, 66, 68–69

roles played in German schools, 19, 20, 24, 28, 51, 53, 87
supported by German churches, 81, 85, 87
Parents and Schools: The 150-year Struggle for Control in American Education, 163n
see also under Cutler, William W.
parochial schools
see under religious schools, American *and* religious schools, German
Pedagogical Anthropology
see under Montessori, Maria
Perchemlides v. Frizzle (1978), 115–116, 181n
see also under home schooling, mentioned
permanent school reform, 147–149
Pestalozzianism, 48, 50–53
Pillars of the Republic
see under Kaestle, Carl F.
PISA 2000: Basiskompetenzen
see under Baumert, Jürgen
PISA tests (Programs for International Student Assessment), 138–141
Platteville, Wisconsin, normal school, 40, 169n
Politics, Markets, and American Schools
see under Chubb, John *and* Moe, Terry
Politik und Schule von der Französischen Revolution bis zur Gegenwart
see under Michael, Berthold and Schepp, Heinz-Hermann
polytechnical high school, 125
see also under Germany, schooling in the German Democratic Republic
Potsdam Agreement, 182n
poverty, its impact on American schools, 95
Power of their Ideas, The
see under Meier, Deborah
Power and the Promise of School Reform
see under Reese, William J.
Private Bildung-Herausforderung für das öffentliche Bildungsmonopol
see under Schlaffke, Winfried *and* Weiß, Reinhold
private school masters, 7–9, 16
private schools, American, 3–4, 18, 72–78, 109–114, 153–156, 181n
see also under Condition of Education 2002 and Alt, Martha Naomi *and* Peter, Katharin
private schools, German, 5, 21, 25, 82–83, 92, 132–135, 155–156, 184n–185n
Private Schools: From the Puritans to the Present
see under Kraushaar, Otto F.
Private Schulen in der deutschen Bidungsgeschichte
see under Deutscher, Eckhard K.
Privatschulen und öffentliches Schulsystem
see under Flitner, Andreas *and* Gallwas, Hans-Ullrich
professionalization of American public education, 63–80
professionalization of German public education, 80–94

Progressive Education Across the Continents, 164n
see also under Lenhart, Volker *and* Röhrs, Hermann
Progressive Education Association, 67, 80
progressive education reforms, 4–5, 66–69, 87–92, 100, 103, 124, 147, 149, 152
Progressivism, American, 45
Prussia
alliance of church and state in, 48–49
Beamten, role of in, 30, 47–48
cameralist and police sciences in, 23
class-based school system in, 22–24
clergy as school supervisors in, 21, 49
compulsory schooling in, 20 -21, 29, 35
education in former Polish territories, 53
education as information and indoctrination in, 53
hybrid system of school government in, 19, 20, 29–30, 56–57
influence on American public education, 6, 35
Jewish schools in, 54–55
Lutheran Church, its role in, 29
public elementary schools in, 55
school bill of 1819, 171n: *see also under* Süvern, Johann Wilhelm von
school deputations in, 20, 21
schools as state institutions in, 20
Section for Culture and Instruction in, 20, 21
teacher seminaries, directors of, 48, 50, 52
teachers, situation of, 51–52
not supported by parents, 52
Prussian Schoolteachers: Profession and Office, 1763–1848
see under LaVopa, Anthony J.
Public Law 89–10 of 1965
see under Elementary and Secondary Education Act of 1965
Public Law 94–142 of 1975
see under Education for all Handicapped Children Act of 1975
Public Law 107–110 of 2001
see under No Child Left Behind Act
Public Money and Parochial Education
see under Lannie, Vincent
public school revival, 6, 11, 14–16, 31
Public School Society of New York
see under schooling in New York City
Public School Voucher Demonstration, A
see under Weiler, Daniel

racial segregation in American schools, 3, 33–34, 95–99
Rand Corporation, 101
rate bills, 10, 13, 36
rational choice philosophers, 150
Reagan revolution, 145
Reden an die deutsche Nation
see under Fichte, Johann Gottlieb

reeducation in post World War II Germany, 6, 121–124
reform pädagogik
  see under progressive education reforms
Reformpädagogik: Eine kritische Dogmengeschichte, 164n
  see under Oelkers, Jürgen
Reframing Educational Policy,
  see under Kahne, Joseph
Regents of the University of the State of New York, 11
regulated free market, 107
regulated school choice, 102, 105
regulated voucher system, 101–102
  see also under vouchers
Reichsgrundschulgesetz of 1920 (National *Grundschul* Law), 84, 122
release time, 79–80
religion in American schools, 1–2, 78–80, 149–150
religion as subject in German schools, 49, 82, 124, 132
Religionsedikt
  see under Wöllner, Johann Christoph von
religious schools, American, 4, 16–19, 42–44, 75–77, 112–117, 152
religious schools, German, 25, 155, 156, 185n
Report on the Condition and Improvement of the Public Schools of Rhode Island
  see under Barnard, Henry
Report on the State of Public Instruction in Prussia
  see under Cousin, Victor
Revolution of 1848, 52
Rights of Man, The
  see under Paine, Thomas
Rise of Massive Resistance, The
  see under Bartley, Numan V.
Rise of the Modern Educational System, The
  see under Müller, Detlef K. *and* Ringer, Fritz, *and* Simon, Brian

Scholarships for Children
  see under Coons, John E. *and* Sugarman, Stephen D.
school choice
  absence of, in Hitler's Germany, 94
  in Cambridge, Massachusetts, 103–104, 106, 180n
  definition of and debate over, 1–6
  endorsed by U.S. Supreme Court, 36
  in Milwaukee, 66, 114–115, 181n
  in New York City, 103–104, 106
  philosophical and ideological bases of, 150–153
  in pre-modern era, 7
  in present-day United States, 156–162
  in Prussia, 30, 53
  and racial segregation, 33–34
  reasons for, 144–147, 149–153
  seen in historical perspective, 147–148
  and utopian communities, 144
  varieties of, 102–105: *see also under* religion in American schools and Just, Richard

school consolidation, 3, 42
school governance
in the Federal Republic of Germany, 130–142
rise of educational administration in the United States, 45–47, 63–66
in the Soviet Occupied Zone of Germany (SBZ)and the German Democratic Republic (DDR), 124–130
in the United States, 63–80
schooling
before and during the American revolution, 8–10
in colonial Massachusetts, 8, 164n
in colonial Virginia, 8
in Milwaukee, 66, 114–115, 181n
see also under school choice in Milwaukee
in New York City, 16–19: *see also under* school choice in New York City
in pre-modern times, 7–8
in Prussia, 5–6, 19–30, 47–58
in state of Delaware, 65
in state of Georgia, 12–13, 34
in state of Massachusetts, 10, 14–16: *see also under* school choice in Cambridge, Massachusetts
in state of Mississippi, 34
in state of New York, 11–13
in state of Wisconsin, 36–44, 114–115
Schooling German Girls and Women
see under Albisetti, James C.
Schule im Vorfeld der Verwaltung
see under Heinemann, Manfred
Schule und Eltern in der Weimarer Republik
see under Wagner-Winterhager, Luise
Schulen im Exil
see under Feidel-Mertz, Hildegard
Schulen in freier Trägerschaft
see under private schools, German
Schullandheime, German, 91–92
Schulpolitik in der pluralistischen Gesellschaft
see under Milberg, Hildegard
Schulpolitik und Schulsystem in der DDR
see under Anweiler, Oskar
Schulrecht und Schulpolitik
see under Heckel, Hans
Schulrechtskunde: Ein Handbuch
see under Heckel, Hans
Schulreform in Preußen, 1809–1919
see under Schweim, Lothar
Schulvielfalt als Verfassungsgebot
see under Jach, Frank-Rüdiger
Schulverfassung and Bürgergesellschaft in Europe
see under Jach, Frank-Rüdiger
Schulwahl
abolished in the Soviel Zone of Occupation and the German Democratic Republic, 122, 129
definition of, 135
limited role of parents in, 82: *see also under* Informationsunterlage des Sekretariats der Kultusministerkonferenz in nineteenth century, 57, 59, 60

in twentieth century, 81, 84, 121–142
Section for Culture and Instruction, 18, 19, 26
see also under schooling in Prussia
segregation
see under racial segregation in American schools
Senatus Academicus
see under schooling in state of Georgia
Separate and Unequal
see under Harlan, Louis R.
separation of state and church, 2, 115–117, 149, 155, 160
Serrano v. Priest, 102, 179n
Slums and Suburbs
see under Conant, James Bryant
Smith-Hughes Act of 1917, 118
social class and schooling, 11, 15, 19, 20, 138–142
Social Ideas of American Educators, The
see under Curti, Merle
Society for Ethical Culture
see under Adler, Felix
Society for Sport and Technic, 128
Soldiers of Light and Love,
see under Jones, Jacqueline
Souter, Justice David H., 159–162
Southern Manifesto
see under southern opposition to racial integration
southern opposition to racial integration, 96–99
see also under racial segregation in American schools
Sozialgeschichte der deutschen Schule
see under Lundgreen, Peter
Staat, Schule Kirche in der Bundesrepublik Deutschland und in Frankreich
see under Starck, Christian
Staat und Erziehung
see under Krautkrämer, Ursula
Staat und Schule oder Staatsschule?
see under Berg, Christa
Staatsbürgerliche Erziehung der deutschen Jugend
see under Kerschensteiner, Georg
standard-based reforms, 118, 119, 144, 146, 153, 157
Standing Conference of the Ministers of Education of the German *Länder*, 136
State and the Non-Public School, 1825–1925, The
see under Jorgenson, Lloyd P.
State Reform School for Boys, Wisconsin, 37
State School for the Blind, Wisconsin, 37
State Superintendent of Public Instruction
Michigan, 13
New York, 13, 27, 31
Wisconsin, 37, 38
State Teachers Association, Wisconsin, 38
Suggestions for Improving Education in Kingdoms
see under Zedlitz, Karl Abraham Freiherr von
systematization of American public schools, 31–47
systematization of German public schools, 47–62

Task Force on Teaching as a Profession
see under *A Nation Prepared*
teacher education, 12, 24, 26, 37, 38
teacher seminars, German, 50, 60
see also under Prussia
teachers
as tutors, 8, 28, 89
testing programs, 118, 146, 153, 157, 158
Thatcher years, 145
There Goes the Neighborhood
see under Reynolds, David R.,
third (non-profit) sector, 153, 184
see also under Weiss, Manfred
Third Plenary Council, 18, 19
Thomas Jefferson and the Development of American Public Education
see under Conant, James Bryant
Time for Results: The Governors' 1991 Report on Education, 180n
township superintendents, 41
Transformation of the School, The
see under Cremin, Lawrence
Traum von der freien Schule, Der
see under de Lorent, Hans-Peter *and* Ullrich, Volker
truant officers, 47

Über die Gleichberechtigung der Realschule
see under Schacht, Ludwig
Union Free High School Law, Wisconsin, 41
United Federation of Teachers, 103
United States Bureau of the Cesus, 175
United States Bureau of Education, 33, 63
United States Commissioner of Education, reports of, 39, 73–75, 175n
U.S. Constitution
establishment clause, 2, 36, 169n, 181n
fifth amendment, 78
first amendment, 2, 36, 79, 115, 116, 149, 159, 169n, 181n
fourteenth amendment, 36, 77, 78, 116
free exercise clause, 2
tenth amendment, 154, 188n
United States Office of Economic Opportunity, 101
United States Supreme Court, 114–117
Board of Education of Central School District No. 1 v. Allen (1968), 188n
Brown v. Board of Education (1954), 3, 96, 117, 178n
Cochran v. Louisiana State Board of Education (1930), 78, 175n
Committee for Public Education and Religious Liberty v. Nyquist (1973), 188n
Everson v. Board of Education of Ewing Township (1947), 36, 117, 159, 169n, 181n, 188n
Farrington v. T. Tokushige (1927), 175n
Green et al. v. County School Board of New Kent County et al. (1968), 178n
Griffin v. County School Board of Prince Edward County (1964), 178n
Keyes v. School District No. 1 (1973), 178n–179n

Lemon v. Kurtzman (1971), 116, 181n, 188n
Levitt v. Committee for Public Education and Religious Liberty (1973), 188n
McCollum v. Board of Education (1948), 79
Meek v. Pittenger (1975), 188n
Mueller v. Allen (1983), 160, 181n, 188n
Poindexter v. Louisiana Financial Assistance Commission (1967), 98, 178n
Pierce v. Society of Sisters (1925), 36, 77, 115, 175n
Plessy v. Ferguson (1896), 33
West Virginia State Board of Education v. Barnette (1943), 175n
Wisconsin v. Yoder (1972), 115, 181n
Wolman v. Walter (1977), 188n
Zelman v. Simmons-Harris (2002), 2, 116, 159–162, 181n, 188n
Zorach v. Clauson (1952), 79, 181n
Utopias and Utopian Thought: see under Manuel, FranK E.

verwaltete Schule, Die
see under Becker, Hellmut
vouchers, 2, 4, 99–102, 105, 106, 114–117, 150, 152, 158, 159

Waldorf Schools, 77, 89, 135, 156
see also under Steiner, Rudolf
Wealth of Nations, The
see under Smith, Adam
Weimar Republic, constitution of, 81–85, 175n–176n, 184n
see also under Germany
welfare state, attack on, 145
Whigs, American, 15–16
White Citizens' Councils
see under southern opposition to racial integration
White House Conference of 1989, 118–119
Whitewater, Wisconsin, normal school, 40
Wilhelm von Humboldt: Werden und Wirken
see under Scurla, Herbert
Wisconsin Journal of Education, 38
Wisconsin Parental Choice Law, 114
Wisconsin Supreme Court
Edgerton case, 44
Wonderful World of Ellwood Patterson Cubberley, 164n
see under Cremin, Lawrence A.

Zur Schulpolitik der Weimarer Republik
see under Führ, Christoph
Zwischen Reform and Reaktion: Preußische Schulpolitik 1806–1859
see under Baumgart, Franzjörg